T0340998

# RESILIENCE ENGINEERING IN PRACTICE, VOLUME 2

# Ashgate Studies in Resilience Engineering

Series Editors

Professor Erik Hollnagel, *Institute of Public Health, University of Southern Denmark, Denmark*

Professor Sidney Dekker, *Key Centre for Ethics, Law, Justice and Governance, Griffith University, Brisbane, Australia*

Dr Christopher P. Nemeth, *Principal Scientist, Cognitive Solutions Division (CSD) of Applied Research Associates, Inc. (ARA), Fairborn, Ohio, USA*

Dr Yushi Fujita, *Technova, Inc., Japan*

Resilience engineering has become a recognized alternative to traditional approaches to safety management. Whereas these have focused on risks and failures as the result of a degradation of normal performance, resilience engineering sees failures and successes as two sides of the same coin – as different outcomes of how people and organizations cope with a complex, underspecified and therefore partly unpredictable environment.

Normal performance requires people and organizations at all times to adjust their activities to meet the current conditions of the workplace, by trading-off efficiency and thoroughness and by making sacrificing decisions. But because information, resources and time are always finite such adjustments will be approximate and consequently performance is variable. Under normal conditions this is of little consequence, but every now and then - and sometimes with a disturbing regularity - the performance variability may combine in unexpected ways and give rise to unwanted outcomes.

The Ashgate Studies in Resilience Engineering series promulgates new methods, principles and experiences that can complement established safety management approaches. It provides invaluable insights and guidance for practitioners and researchers alike in all safety-critical domains. While the Studies pertain to all complex systems they are of particular interest to high-hazard sectors such as aviation, ground transportation, the military, energy production and distribution, and healthcare.

# Resilience Engineering in Practice, Volume 2

## Becoming Resilient

Edited by

CHRISTOPHER P. NEMETH
*Applied Research Associates, Inc., USA*

ERIK HOLLNAGEL
*University of Southern Denmark, Denmark*

ASHGATE

© Christopher P. Nemeth and Erik Hollnagel 2014

All rights reserved. No part of this publication may be reproduced, stored in a retrieval system or transmitted in any form or by any means, electronic, mechanical, photocopying, recording or otherwise without the prior permission of the publisher.

Christopher P. Nemeth and Erik Hollnagel have asserted their rights under the Copyright, Designs and Patents Act, 1988, to be identified as the editors of this work.

Published by
Ashgate Publishing Limited
Wey Court East
Union Road
Farnham
Surrey, GU9 7PT
England

Ashgate Publishing Company
110 Cherry Street
Suite 3-1
Burlington, VT 05401-3818
USA

www.ashgate.com

ISBN: 9781472425157 (hbk)
ISBN: 9781472425164 (ebk – PDF)
ISBN: 9781472425171 (ebk – ePUB)

**British Library Cataloguing in Publication Data**
A catalogue record for this book is available from the British Library

**The Library of Congress data has been applied for.**

Printed in the United Kingdom by Henry Ling Limited,
at the Dorset Press, Dorchester, DT1 1HD

# Contents

# List of Figures

# List of Tables

# Preface: Seeking Resilience

Christopher Nemeth

Becoming Resilient is the second text in the Ashgate series "Resilience Engineering in Practice (REiP)." While Ashgate Publishing's "Resilience Engineering Perspectives" series has explored what the field of resilience engineering (RE) is, REiP take the practical approach to RE. The chapters in this text seek answers to the challenging questions that are posed by applying concepts in prior texts to actual problems. Their reports show that while the first successful steps have been made, there is still a lot to do in order to develop RE from an initial concept into an approach that will change the way systems are developed and operated.

## Opportunities for Engineering Practice

The creation of systems that are ready to evolve in response to unforeseen conditions poses a challenge to also develop a new way to think about design and engineering. Designers and engineers typically develop systems, and engineers are entrusted with ensuring systems are built to operate according to requirements. But new approaches such as RE call for new abilities. What professional abilities are needed to create systems that have the resilient characteristics that chapters in this text describe? What skills and opportunities will engineers need in order to develop systems that can adapt to meet unforeseen demand?

For years, radio towers used a strong base to withstand the effects of high winds. Their rigid design, though, limited how high they could be built. The invention of slender radio masts, held in place by guy wires, made taller towers possible by allowing the structure to move in response to the wind instead of standing rigidly against it. Engineering practice faces a similar transition.

Engineers have traditionally sought ways to maintain sufficient margins to assure safe performance. In the process they have developed a resistance to sources of variability that could affect those margins. This may fit well-bounded stable domains, where sources of variability are fairly well known. However, poorly bounded and ill-behaved domains are increasing in number and importance. Domains such as these routinely make demands that can only be met by socio-technical systems (Hollnagel and Woods, 2005), which are the goal-directed collaborative assembly of people, hardware and software. In these systems, their elements operate collectively, not individually. Woods (2000) referred to the interaction of all system elements as "agent-environment mutuality." Their performance and interaction provide outcome behavior, and the data that can be gathered on their performance can be compared against requirements.

Engineering is the application of science and mathematics "by which the properties of matter and the sources of energy in nature are made useful to people" (Merriam Webster, 2013). Systems engineering (SE), which has a significant role in RE, integrates multiple elements into a whole that is intended to serve a useful purpose. SE "... is an interdisciplinary approach and means to enable the realization of successful systems" that "focuses on defining customer needs and required functionality early in the development cycle, documenting requirements, then proceeding with design synthesis and system validation while considering the complete problem: operations, performance, test, manufacturing, cost and schedule, training and support, disposal" (INCOSE, 2013). To do this, SE "... integrates all the disciplines and specialty groups into a team effort forming a structured development process that proceeds from concept to production to operation" and "considers both the business and the technical needs of all customers with the goal of providing a quality product that meets the user needs" (INCOSE 2013). The process assembles elements into a coherent whole, but how does that whole operate? How does it respond to demands? What happens when it reaches the upper bounds of its ability to withstand a challenge? Answers to these and other challenges will come from new approaches by those who develop these systems.

## A Resilient Outlook

In order to develop systems that function in a resilient manner, engineering has the opportunity to grow in a number of different ways. Effective engineering ensures positive outcomes from a system's performance. To make resilience routine, engineers might foresee what may go wrong, develop new tools, and use good design to model adaptive solutions. Here are a number of initiatives that can make that intention reality.

*Reconceptualize*. Mapping all possible interdependencies among system elements is too difficult, because many of them are hidden. Instead, approach the design problem at a higher level that allows for anticipation, as well as needed change, from simple reconfiguration to more complex needs to expand and adapt. In addition to centralized command architectures and flat architectures, consider multi-role, multi-echelon networks.

*Study what goes right*. In contrast to the traditional safety focus on failures, resilience engineering emphasizes the importance of focusing on what works—on what goes right (Hollnagel, 2014). This requires us to pay attention to that which we routinely neglect simply because it "just happens." Learning to do so is not very difficult, since it is a question of changing what we look for rather than to dig deeper. Contrary to traditional safety thinking, breadth is more important than depth.

*Cultivate requisite imagination*. The measure of an organization's success is the ability to anticipate changes in risk before failure and loss occur; to create foresight (Woods, 2000). Adamski and Westrum (2003) describe requisite imagination as the ability to foresee what might go wrong, and maintain a questioning attitude throughout the development process. Aspects of practice they consider essential to this trait include thoroughly defining the task to be performed, identifying organizational constraints, matching the world of the system designer with that of the system user, considering the operational environment and the domain where work will be performed, surveying past failures, using controls appropriate to the tasks to be performed, accounting for potential erroneous actions, and taking conventions and constraints into account. This is no small job, and it calls for further research to understand how to support the tasks that this kind of foresight requires.

*Develop new tools to develop and operate systems.* System engineering tools and knowledge management tools must incorporate human and organizational risk. Based on empirical evidence, develop ways to control or manage a system's ability to adapt. This includes developing ways for a system to monitor its own adaptive capacity so that it can make changes in anticipation of future opportunities or disruptions. Provide feedback, using knowledge about operations to identify conditions when to launch analyses of key system features.

Resilience engineering includes operational oversight, which is typically termed "management." However this type of oversight means more than what management normally implies. RE is similar to management because it requires the system to be self-aware; able to reflect on how well it has adapted. It is different because it goes beyond operations to include research and development to ensure it has the traits that make it adaptable in the first place.

Management that correctly understands the operations of any system will also be likely to correctly estimate how well its strategies will work when unforeseen challenges occur. While management points of view influence how systems are to be configured, they may not reflect the realities of operational demands. Managers who don't understand the operator's point of view at the sharp end can miss the demands and constraints operators face. Their well-intentioned efforts that do not reflect an understanding of sharp end issues can produce both doctrinal and technological surprise. For example, healthcare organizations experience such misunderstandings that include software development cost overruns, alert overload, mode errors that include operating on the wrong patient or patient site, and a general increase in the number of shortcuts needed to compensate for cumbersome and inflexible technology.

Management's part in this includes balancing production pressure with protection from loss. Foster a culture that encourages reporting. Respond with repair or authentic reform when circumstances call for it. Enable front-line supervisors to make important decisions in order to be aware of and act on problems as evidence begins to develop. Understand operations well enough to know when they are encroaching on safety boundaries.

*Create ways to monitor the development and occurrence of unforeseen situations.* Complex systems are dynamic and need means to not only monitor performance but also make deliberate adjustments to anticipate, and respond to, unforeseen situations. Operators know how they engage and deal with these. Front-line workers have identified 8–12 workplace and task factors that can make work difficult, including interfaces with other groups, input information that is partial or missing, and staff and resource shortages (Reason, 1997). Wreathall (2001; 2006) and Wreathall and Merritt (2003) reviewed sets of indicators that map onto aspects of resilience. Such measures point to the onset of problems in normal work practices as pressures grow. They also reveal where workers develop adjustments to compensate for that. Management is usually unaware of changing demands or of the need for workplace adjustments. These indicators are chosen to reveal circumstances and can also reveal situations management may not know about, and current plans may not be adequate to handle changing demands.

*Develop tools to signal how to make production vs safety tradeoffs and sacrifice decisions.* Enable an organization to know when to relax production goals in order to reduce the risk of coming too close to safety boundaries, even under uncertain conditions. Learn how organizations consider and make these decisions, as well as what is needed to support them.

*Cultivate ways to visualize and foresee side effects.* Develop means to show how systems adjust their performance to handle unexpected situations. Show how pressure from other units or echelons affects a particular portion of a system. Take interactions with other systems into account, and be aware of the implications that interactions present, such as cascading effects.

*Promote and use good design.* The development of well-considered prototypes makes it possible to evaluate how, and how well, solutions can adapt. Norman (2011) contends that "Good design can help tame complexity, to relish its depth, richness, not by making things less complex—for the complexity is required—but by managing the complexity." Good design can be used to model discoveries of adaptations to change and uncertainty. The prototypes that result offer compelling evidence of a feasible future that others can understand and evaluate.

*Acknowledge and manage variability.* New configurations introduce uncertainty, and bounding a problem to exclude uncertainty does not eliminate it. Embrace non-linear approaches to explore how systems and networks adapt to change and disruption.

Like RE, each of these opportunities challenges the imagination to move professional practice from what is known to what it can, and needs to, become.

## Reading Guide

A brief comment follows each of the chapters to invite the reader's attention to key points that connect with the book's theme and occasionally describe how the chapter relates to the one that follows.

Each chapter examines the need for RE in actual settings, including healthcare, nuclear power, aviation, railway tunnels, construction, and disaster recovery. The chapters explore practical issues that will need to be resolved, and new approaches that will be needed to make RE feasible. Understand how systems work in reality. Translate a system description into a prescription to improve its adaptive ability. Negotiate the differences between work-as-imagined and work-as-done, and learn from the experience. Learn and anticipate as a way to cope with fundamental surprise. Pay attention to more subtle safety indicators such as process safety and organizational hazards. Analyze adaptation as a way to improve system monitoring and systemic learning. Understand how interplay among multiple levels and actors can influence socio-technical systems. Translate team training from individual to a distributed cognition approach that recognizes tasks are variable. Gain and retain a new perspective to notice what couldn't be seen before and, once it is noticed, compels one to act. Triangulate incomplete or ambiguous readings from multiple sources during an event, and determine what gaps in these readings say about the role of human cognition in achieving resilience.

The concept of adaptive systems refers to a result of the way systems perform. "Becoming Resilient" implies that what we describe in these pages is a process. As each chapter shows, the process can include new approaches to methods, organizational structures, and work processes.

System concepts take time to evolve, and the development of RE will also take time to develop the science, measures, and means that other approaches already have in place.

## Acknowledgement

The author is grateful to David Woods, John Wreathall, and Erik Hollnagel for their insightful comments during the development of this preface.

# Notes on Contributors

**Marcus Abrahamsson**, PhD, is Head of the Division of Risk Management and Societal Safety at Lund University. His research interests include the design of methods for risk, vulnerability, and capacity assessments for enhanced resilience in various contexts. Marcus combines his academic and educational career with work in international development cooperation focused on disaster risk management.

**Per Becker** is Associate Professor and Director of Centre for Societal Resilience, Lund University. He has combined research and education with a career in humanitarian assistance and international development cooperation focused on Disaster Risk Reduction, Recovery, and Conflict Management. Per has extensive field experience, and is still involved with national authorities and international organizations active in promoting a safe and sustainable world, most recently as Regional Disaster Risk Management Coordinator at the IFRC regional office in Dakar. Per is interested in transdisciplinary research of sustainability and social change, of what makes society resilient to disturbances, disruptions and disasters, and of capacity development as an intentional tool for creating and maintaining such resilience. Per is also interested in researching the role of vulnerability in creating and maintaining public support for conflict.

**Johan Bergström**, PhD, is Associate Professor at Lund University, Division for Risk Management and Societal Safety (Sweden). Johan's current research is focusing on the notion of Societal Resilience; one which is currently being implemented in regional and national policies all over the world. Johan's chapter however mainly reflects on research conducted during the time that he was still a PhD candidate, studying organizational resilience in escalating situations.

**Matthieu Branlat**, PhD, is a Research Scientist at 361 Interactive, LLC in Springboro, OH. He obtained a PhD in Cognitive Systems Engineering from the Ohio State University in 2011. His research interests include resilience engineering and safety, decision-making and collaborative work. His projects are conducted in domains such as urban firefighting, military rescue, industrial maintenance, intelligence analysis, cyber security, and patient safety.

**Alexander Cedergren,** PhD, is a Researcher at the Division of Risk Management and Societal Safety at Lund University, Sweden. He is affiliated to Lund University Centre for Risk Assessment and Management (LUCRAM) and Lund University Centre for Societal Resilience. His main research interests include resilience engineering, risk governance, accident investigation, and analysis of interdependencies and vulnerability of critical infrastructures.

**Nicklas Dahlström** is Human Factors Manager at Emirates Airline and has been with the airline since 2007. In this position he has overseen CRM training in a rapidly expanding airline and also been part of efforts to integrate Human Factors in the organization. Nicklas was previously a researcher and instructor at Lund University School of Aviation in Sweden, working mainly on projects related to safety and Human Factors in aviation as well as in other areas, such as maritime transportation, nuclear industry, and health care. His research areas in aviation have been mental workload, training, and simulation and he has written research articles and book chapters on Human Factors and CRM as well as delivered invited presentations, lectures, and training in more than a dozen different countries.

**Camila Campos Gómez Famá** is Professor at the Instituto Federal de Educação, Ciência e Tecnologia da Paraíba (IFPB), in Brazil. She is a civil engineer (2007) and has an MSc in construction management (2010). Her main research interests are related to construction safety and entrepreneurship.

**Carlos Torres Formoso** is Professor in Construction Management at the Federal University of Rio Grande do Sul (UFRGS), Brazil. He has a degree in Civil Engineering (UFRGS, 1986), an MSc in Construction Management (UFRGS, 1986), and a PhD (University of Salford, 1991). He was a Visiting Scholar at the University of California at Berkeley (1999–2000), and a Visiting Professor at the University of Salford, UK (2011). His main research interests are production planning and control, lean production, performance measurement, safety management, social housing, and value management.

**Nicolas Herchin** MPhil, is research engineer and project manager in the research and innovation division of GDF SUEZ, in Paris. After graduating from Cambridge University, UK, in Industrial Systems, Manufacturing and Management, he is now leading since 2009 a project in the field of Human and Organizational Factors of Safety. As such, he works tightly with the Group's gas infrastructure affiliates (including transportation, storage and LNG terminals) on improving safety aspects, relying on strong partnerships with universities and French institutes in the fields of resilience engineering, safety culture, or high reliability organizations and developing tailor-made tools and approaches for the energy industry.

**Éder Henriqson** is Associate Professor at the School of Aeronautical Science at Pontifícia Universidade Católica do Rio Grande do Sul (Brazil) and Affiliated Professor at Lund University (Sweden). His research interests are organizational safety, resilience engineering, accident investigation, and cognitive systems engineering.

**Erik Hollnagel**, PhD, is Professor at the University of Southern Denmark, Chief Consultant at the Center for Quality Improvement, Region of Southern Denmark, Visiting Professorial Fellow at the University of New South Wales (Australia), and Professor Emeritus at University of Linköping (Sweden). He has worked at universities, research centres, and industries in several countries since 1971, with problems from several domains, including nuclear power generation, aerospace and aviation, air traffic management, software engineering, healthcare, and land-based traffic. His professional interests include industrial safety, resilience engineering, accident investigation, systems thinking, and cognitive systems engineering. He has published more than 250 papers and authored or edited 22 books. Some of the most recent titles include *Safety-I and Safety-II* (Ashgate, 2014), *Resilient Health Care* (Ashgate, 2013), and *The Functional Resonance Analysis Method* (Ashgate, 2012). Erik is also Editor-in-Chief of Ashgate Studies in Resilience Engineering.

**Masaharu Kitamura** is President of Research Institute for Technology Management Strategy which he founded in 2012. Previously he served as a faculty member of Tohoku University, Department of Nuclear Engineering for 36 years and now he is Emeritus Professor at Tohoku University. His professional areas include instrumentation and control of nuclear power plants, Human Factors and organizational safety in nuclear and general industries, and ethics in engineering. He is also active in the areas of public dialogue on nuclear risk and resilience engineering.

**Akinori Komatsubara** is Professor at the Department of Industrial and Management Systems Engineering at the School of Science and Engineering of Waseda University in Tokyo (Japan). He has a PhD in Industrial Engineering and Human Computer Interaction Studies. He has studied in the area of industrial safety, human performance enhancement, cognitive usability studies, non-technical skills, and their management systems. He has also worked in Japan for several airlines, railways, and nuclear industries as their safety advisor.

**Jean-christophe Le Coze** is a safety scientist with an interdisciplinary background, including engineering and social sciences. He works at INERIS, the French National Institute in Environmental Safety. His activities combine ethnographic studies and action research programs in various safety critical systems with an empirical, theoretical, historical and epistemological orientation. Outcomes of these researches have been regularly published in the past ten years.

**Elizabeth Lay** is Director of Human Performance for Calpine Corporation, Houston Texas, US. Calpine is the United States' largest independent power producer based on megawatts generated, with more than 90 plants in the US and Canada. She has written papers and contributed to several books on Resilience Engineering. She has worked in the energy industry in the domain of operations risk management for 10 years. She has a Bachelor of Mechanical Engineering (BSME) degree and graduate certificate in Cognitive Science.

**Jonas Lundberg**, PhD, is Senior Lecturer in Information Design at the Department of Science and Technology, Linköping University, Sweden. He obtained his PhD in Computer Science from Linköping University in 2005. His research concerns information design in high stakes domains, and the fields of resilience engineering, cognitive systems engineering, and human work interaction design.

**David Mendonça**, PhD, is Associate Professor in the Industrial and Systems Engineering Department at Rensselaer Polytechnic Institute. His research examines the cognitive processes underlying individual and group decision-making in high stakes, time-pressured conditions, particularly during post-disaster emergency response. This research has employed data collected in laboratory, field, and archival settings, leading to statistical and computational models, as well as to systems that support cognition and learning in the target domains. His work has been supported by numerous grants from the US National Science Foundation. He received his BA from University of Massachusetts, MS from Carnegie Mellon University, and PhD from Rensselaer Polytechnic Institute. He has been a Visiting Scholar at Delft University of Technology (The Netherlands) and the University of Lisbon (Portugal).

**Christopher Nemeth**, PhD, is a Principal Scientist III and Group Leader for Cognitive Systems Engineering at Cognitive Solutions Division of Applied Research Associates, Inc. His design and human factors consulting practice and his corporate career have encompassed a variety of application areas, including healthcare, transportation, and manufacturing. As a consultant, he has performed human factors analysis and product development, and served as an expert witness in litigation related to human performance. His research interests include technical work in complex high stakes settings, research methods in individual and distributed cognition, and understanding how information technology erodes or enhances system resilience. He has served as a committee member of the National Academy of Sciences, is widely published in technical proceedings and journals, and his books include *Human Factors Methods for Design* (Taylor and Francis/CRC Press, 2004), as well as Ashgate Publishing texts *Improving Healthcare Team Communication* (2008), and Resilience Engineering Perspectives Series Volume One—*Remaining Sensitive to the Possibility of Failure* (2008) and Volume Two—*Preparation and Restoration* (2009).

**Amy Rankin** is a PhD student in Cognitive Systems at the Department of Computer and Information Systems at Linköping University. She has a Fil. lic. in Cognitive Systems (2013) from Linköping University and her research interests include resilience engineering, cognitive systems engineering, safety culture, and human factors.

**Tarcisio Abreu Saurin, Dr,** is Professor at the Industrial Engineering Department of the Federal University of Rio Grande do Sul (UFRGS) in Porto Alegre, Brazil. His main research interests are related to safety management in complex systems, resilience engineering, lean manufacturing, and production management. He has worked as a coordinator and/or researcher in funded projects related to those topics in several sectors, especially construction, electricity distribution and generation, manufacturing, aviation, and healthcare. The results of his studies have been published in a number of journals and conferences.

**Henrik Tehler** is Associate Professor at the Division of Risk Management and Societal Safety, Lund University. His professional interests include risk governance, disaster risk reduction, societal safety, resilience engineering, and decision-making.

**Robert L. Wears**, MD, PhD, is Professor in the Department of Emergency Medicine at the University of Florida, and Visiting Professor in the Clinical Safety Research Unit at Imperial College London. His research interests include technical work studies, resilience engineering, and patient safety as a social movement. His authored or co-edited books include *Patient Safety in Emergency Medicine* and *Resilient Health Care*. A new title, co-edited with Erik Hollnagel and Jeffrey Braithwaite, entitled *Resilience in Everyday Clinical Work*, is expected in 2014.

**L. Kendall Webb**, MD, is Associate Chief Medical Information Officer at UF Health Systems as well as Vice Dean of Medical Informatics and Assistant Professor of Emergency Medicine and Pediatric Emergency Medicine for the University of Florida in Jacksonville, Florida. Previously, she was a Senior Software and Systems Engineer over a 10-year career with Raytheon/E-Systems in the Washington, DC. Her areas of expertise include full life-cycle development with implementation and optimization of software and systems applications—most recently electronic health record systems, usability, patient safety, resilience, process engineering, effective communication, and quality. She has created multiple interdepartmental curriculums related to the Emergency Department and implemented upgrades to core hospital processes.

**Rogier Woltjer**, PhD, is Senior Scientist at the Swedish Defence Research Agency (FOI), Division of Information and Aeronautical Systems. He also works as part-time Assistant Professor at the Department of Computer and Information Science of Linköping University, Sweden. He obtained a PhD in Cognitive Systems there in 2009. His research and work with industry has addressed training, decision support, command and control, risk analysis, incident investigation, and safety and security management. Application domains include air traffic management, aviation, and emergency and crisis management.

# Reviews for *Resilience Engineering in Practice, Volume 2*

Resilience engineering is becoming the new paradigm for conceptualising safety in complex systems. Adopting a proactive, socio-technical, systems-based approach to safety is becoming de rigueur in high jeopardy-risk endeavours. With contributions from international contributors with a wealth of experience, this book is essential reading for both practitioners and researchers who are interested in the application of resilience engineering principles in practice.

Don Harris, Coventry University, UK

There is no doubt that resilience is a core requirement for handling the risks of today´s and tomorrow's ever more complex socio-technical systems. This book provides the perfect mix of conceptual discussion and practical application to guide academics and practitioners towards designing more resilient systems.

Gudela Grote, ETH Zürich, Switzerland

# Chapter 1
# An Emergent Means to Assurgent Ends: Societal Resilience for Safety and Sustainability

Per Becker, Marcus Abrahamsson and Henrik Tehler

Societal safety and sustainability are key challenges in our complex and dynamic world, causing growth in interest of applying the concept of resilience in broader societal contexts. This chapter presents a concept of societal resilience that builds on established theory of Resilience Engineering and operationalises the concept by presenting its purpose, required functions and a way to identify and analyse the complex network of actual forms that together achieve these functions in society. The framework for analysing societal resilience is then tested in practice with interesting results. Although the framework has challenges and limitations, the Resilience Engineering approach to societal resilience seems to be both a conceptually and pragmatically fruitful path to follow.

## Introduction

Contemporary society seems preoccupied with the notion of risk and recent examples of calamity have given rise to growing public discontent with the performance of present risk management institutions (Renn, 2008, p. 1). The safety and sustainability of society is thus increasingly becoming the centre of attention of policymakers from various administrative levels and countries around the world (for example, OECD, 2003; Raco, 2007). Advancing safety and sustainability is challenging in this context, as there are multiple stakeholders to involve (Haimes, 1998, p. 104; Renn, 2008, pp. 8–9), values to consider (Belton and Stewart, 2002), and stresses to include (Kates et al., 2001, p. 641). On top of this lies the multitude of factors and processes contributing to the susceptibility of what stakeholders' value to be the impact of each stress (Wisner et al., 2004, pp. 49–84; Coppola, 2007, pp. 146–61).

The real challenge of societal safety and sustainability is however not the number of elements to include, but the complexity and non-linearity of relations between these elements (Yates, 1978, p. R201), separating cause and effect in both space and time (Senge, 2006, p. 71). Unfortunately, in efforts to promote safety and sustainable development, stakeholders often reduce problems into parts that fit functional sectors, organisational mandates and academic disciplines (Fordham, 2007). This is likely to be a major weakness as it clouds the bigger picture of risk (Hale and Heijer, 2006, p. 139) and is further complicated by various processes of change increasing the dynamic nature of our world, for example, globalisation (Beck, 1999), demographic and socio-economic processes (Wisner et al., 2004), environmental degradation (Geist and Lambin, 2004), the increasing complexity of modern society (Perrow, 1999b) and climate change (Elsner et al., 2008). It is in this context that Resilience Engineering may offer a conceptual framework to build on for meeting the challenges of societal safety and sustainability in the 21st century and beyond.

The purpose of this chapter is to present a framework for addressing challenges to the safety and sustainability of societies, by defining and operationalising a concept of societal resilience. The chapter also presents examples from applications of the framework in different contexts.

## A Concept of Societal Resilience

Resilience Engineering has been paramount in demonstrating that the main challenge for safety is to recognise dynamic complexity and non-linear interdependencies in the system in question (for example, Hollnagel, 2006, pp. 14–17). Similarly, Sustainability

Science has been equally paramount in demonstrating the same challenge for sustainability (for example, Kates et al., 2001). While Resilience Engineering appears to have generally been focusing on socio-technical systems (for example, Cook and Nemeth, 2006, p. 206; Leveson et al., 2006, p. 96), Sustainability Science has often approached our world as a complex human-environment system (for example, Turner et al., 2003; Haque and Etkin, 2007). Regardless of type of system, destructive courses of events that threaten safety and sustainability are, in both views, not results of linear chains of events, like dominos falling on each other (Hollnagel, 2006, pp. 10–12), but are instead non-linear phenomena that emerge within the complex systems themselves (Perrow 1999a; Hollnagel, 2006, p. 12). Such destructive courses of events are thus not discrete, unfortunate and detached from ordinary societal processes, but intrinsic products of everyday human-environment relations over time (Hewitt, 1983, p. 25; Oliver-Smith, 1999), and rooted in the same complex system that supplies human beings with opportunities (Haque and Etkin, 2007).

The concept of resilience has a wide range of definitions, developed in scientific disciplines spanning from engineering to psychology. Many of these definitions describe resilience as ability to 'bounce back' to a single equilibrium (for example, Cohen et al., 2011), as a measure of robustness or buffering capacity before a disturbance forces a system from one stable equilibrium to another (Berkes and Folke, 1998) or as ability to adapt in reaction to a disturbance (Pendall et al., 2010). Although many of these definitions are useful for their intended purposes, human-environment systems are adaptive and entail human beings with the ability not only to react to disturbances but also to anticipate and learn from them. For instance, a bicycle lane with a curb crossing would not be considered a particularly resilient system, even if it involved a qualified nurse and bike repairman ready to aid the unfortunate passerby to get back into the saddle again. It would be more resilient if the curb is removed, either before the foreseeable accident happens (anticipation), or after the first incident (learning).

To advance societal safety and sustainability, it is crucial to approach society as a complex human- environment system, and its level of safety and sustainability is determined by internal attributes. Societal resilience is in this sense an emergent property of such a system in the same way as Pariès' (2006) organisational resilience of complex organisations. To better grasp this emergent property, Rasmussen (1985) suggests to structure systems in a functional hierarchy from purpose, through increasingly concrete levels of function, to the observable physical forms of the system contributing in the real world to fulfil the functions and meet its purpose. However, it is important to note that resilience is not a linear outcome of these functions, but an emergent property of the system that is connected in complex ways to the ability of the system to perform.

In the context of societal resilience, the overarching purpose of the human-environment system under study is to protect what human beings value, now and in the future. Hollnagel's (2009) four cornerstones of resilience form a comprehensive foundation for the functions fulfilling that purpose. Although his framework is compelling, with its focus on anticipation, monitoring, responding and learning (Ibid.), it needs some minor alterations to suit the broader societal context. More specifically, to increase the usefulness of it in practice we suggest the introduction of a set of general functions on a lower level of abstraction (see Rasmussen, 1985) that will facilitate the assessment of societal resilience. It is important to note that although Hollnagel's four pillars, and the associated abstract functions in our approach, are complete, the more concrete generalised functions used in this chapter are not. There may in other words be other generalised functions that are useful in other contexts, and the ones presented may be divided in other ways.

Moreover, we also suggest that these generalised functions could be divided into proactive and reactive ones. A function is here defined as proactive if it has an ex ante focus, that is, it focuses on something that has not yet happened. A reactive function, on the other hand, has an ex post focus, that is, it focuses on an actual event that has already taken place. This division into groups of reactive and proactive functions is useful in the present context since it emphasises that societal resilience is not only about 'bouncing back' from a disturbance, but also about adapting the system beforehand and learning from previous events. Furthermore, the division into two sets of functions also allows us to study the interaction between the two types, for example between the preparedness and response functions (see below), and it allows us to specifically address the different contextual factors that influence the performance of the proactive functions compared to the reactive ones. Examples of contextual factors that usually exist to a greater extent for reactive functions than proactive ones include high time pressure, large stakes and rapidly changing conditions.

Starting with the first cornerstone of Hollnagel's framework, the function of anticipation, we suggest that a more concrete generalised function in a societal context is the function of Risk assessment. We agree with Hollnagel when stating that methods for risk assessment that focus on linear combinations of discrete events may fail to sufficiently represent risk as they fail to take into account the complexity of our world (Ibid., pp. 125–7). However, this is not a general attribute of risk assessment per se, and there are methods that to a greater extent incorporate such complexity (Haimes, 2004; Petersen and Johansson, 2008). There is obviously no such thing as a perfect method for risk assessment, but, as Hollnagel admits, 'a truly resilient organization realizes the need at least to do something' (Hollnagel, 2009, p. 127). A related but perhaps less contentious way of anticipation is Forecasting, for example, weather forecasts, river flow as a result of potential rainfall, ocean waves if a storm grows stronger and so on. Both Risk Assessment and Forecasting are proactive functions.

The second cornerstone emphasises the need to monitor specific predefined indicators of potential problems (Ibid., pp. 124–5), for example actual river flow, number of cholera cases in the area and so on. Hollnagel's concept of monitoring covers in other words what 'is or could be a threat in the near term' (Ibid., p. 120), but not functions that are vital when the system is already in a specific disastrous event. In such a situation the system needs a function to recognise what impact that event has on the system. It is thus suggested that the second cornerstone of resilience is modified and called recognising, covering the generalised functions of Monitoring and Impact assessment. The latter clearly being reactive, while the former being potentially both proactive and reactive depending on how the set value for the indicator that is being monitored is defined in relation to what constitutes a real crisis (Figure 1.1). Impact assessment is in other words not an abstract function corresponding to Hollnagel's cornerstones in itself, but an addition to his framework. The name of the associated abstract function in our approach is thus changed from Monitoring to Recognising to indicate that.

The third cornerstone accentuates the importance to be able to adapt the system in different ways based on what is anticipated to have a potential to become a problem in the future, what is recognised as critical or soon to be critical in the current situation or what is learnt to be a problem from experience. Hollnagel (2009) calls this responding, but includes adaptations to respond to and recover from specific events, as well as different ways to prevent/mitigate or prepare for an adverse event. As responding in the broader societal context connotes only the reactive response to a disaster situation, the name of the cornerstone is altered to Adapting. Generalised functions in a societal context corresponding to Adapting are Prevention and mitigation, Preparedness, Response and Recovery, where the two former are proactive and the two latter are reactive functions.

Hollnagel's fourth cornerstone is Learning, as he clearly states that a 'resilient system must be able to learn from experience' (Ibid., p. 127). What failed in a specific disastrous event, as well as who is to blame for it, is not the focus here. Learning should instead be a continuous planned process focused on how the system functions, links between causes and effects, its interdependencies and so on (Ibid., pp. 129–30). In a societal context, learning from disturbances is often associated with evaluations of what happened and how various actors responded to the event in question. We call the generalised function Evaluation, which can be both proactive and reactive as it is not only possible to learn from what has happened in reality but also from counterfactual scenarios (Abrahamsson et al., 2010). It should however be noted that the actual learning in a system is very much dependent on the feedback loops from Evaluation to the other generalised functions (Figure 1.1), requiring that changes are made based on this input. There are many other aspects of learning that are not captured by the generalised function of Evaluation, but, as stated earlier, the generalised functions presented in this chapter are a selection for a specific societal context.

We argue that societal resilience is an emergent property determined by society's ability to anticipate, recognise, adapt to and learn from variations, changes, disturbances, disruptions and disasters that may cause harm to what human beings value. Above, we have suggested how these four abstract functions can be transformed into generalised functions that are more concrete and provides more guidance on how to identify them in a societal context.

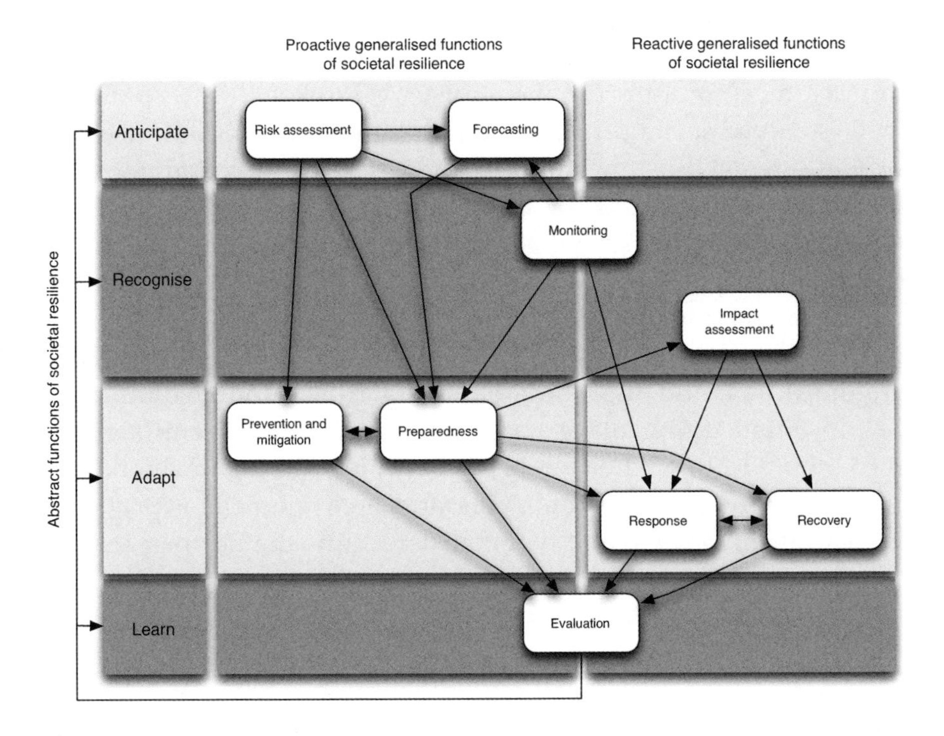


**Figure 1.1 The abstract and generalised functions of societal resilience**

## Operationalising Societal Resilience

To meet the stated purpose of societal resilience, the system under study must have sufficient capacities for the abstract functions of anticipation, recognising, adapting and learning, which can be further specified by the generalised functions of risk assessment, forecasting, monitoring, impact assessment, prevention/mitigation, preparedness, response, recovery and evaluation (Figure 1.1). It is important to note that just as there may be many ways to describe any system (Ulrich, 2000, pp. 251–3), there may be different ways of concretising the four overall abstract functions for societal resilience. However, it is also important to note that there are dependencies between these functions making the functioning of one dependent on the output of the functioning of others, for example, for the public to be able to undertake the preparedness measure to take shelter for a coming cyclone necessitates warning information from forecasting or monitoring the weather (Figure 1.1).

To analyse societal resilience in a particular context, it is not enough to establish what functions that are needed to meet the purpose of protecting what human beings value. For that we need to focus on what Rasmussen (1985) calls form, on the observable aspects of the real world that together constitute the required functions of the system. These forms include: (A) legal and institutional frameworks; (B) systems of organisations; (C) organisations; and (D) human and material resources, but may be presented under other headings (for example, Schulz et al., 2005, pp. 32–50; CADRI, 2011). Analysing societal resilience is, in other words, about identifying and analysing the aspects, on these multiple levels, that together determine the performance of the functions of societal resilience.

To be able to identify and analyse vital aspects for the societal resilience of a particular system, a set of 22 guiding questions have been developed that need to be answered for each of the nine generalised functions. The questions extend over all four levels (A–D), and are presented in Table 1.1 Useful approaches to finding answers to these guiding questions include conducting focus groups with relevant stakeholders on various administrative levels, interviews with key informants and document studies.

These ideas have been applied with interesting results in capacity assessments of the system for disaster risk management and climate change adaptation in Botswana and Tanzania. The studies were done for MSB, a Swedish governmental humanitarian and development cooperation agency, and together with the Botswana National Disaster Management Office (NDMO) and the Tanzania Disaster Management Department (DMD) respectively.

The purpose of the studies was to generate input to the formulation of capacity development projects towards strengthened resilience in the two countries. In some sense as a contrast to traditional capacity development projects in the field of disaster risk management, which often focuses on some specific aspect of disaster risk management capacity, the intention here was to produce a comprehensive and holistic overview of disaster risk management capacities at national, district and local level in the two countries. This was to able to design a set of capacity development interventions targeting the most important functions for societal resilience in the two contexts, and to do this at the most suitable level while considering the vast interdependencies between the different functions. The broad scope described above made the framework presented in this chapter suitable as the methodological underpinning of the studies.

To exemplify the potential means for finding the answers to the guiding questions in Table 1.1, in the Botswana case the research team conducted focus groups with stakeholders involved in disaster risk management on national, district and local level, as well as interviews with a number of key stakeholders. All in all, 36 stakeholders were involved in the process, spanning from the Botswana Defence Force to a Village Development Committee, and from a Deputy Paramount Chief to the Department of Water Affairs. The research team also studied legislation and policy documents relevant for disaster risk management in Botswana. Similarly, in the case of Tanzania, the research team conducted focus groups and interviews with stakeholders involved in disaster risk management on national, regional, district and local levels. This time, 55 stakeholders were involved in the process, spanning from UN agencies to Village Committees, and from the Ministry of Education & Vocational Training to a District Social Welfare Department. Again, the research team also studied legislation and policy documents relevant for disaster risk management in Tanzania.

**Table 1.1 Examples of guiding questions for capacity assessment of systems for disaster risk management and climate change adaptation**

| | Levels of factors determining capacity | | | |
|---|---|---|---|---|
| **Functions** | **A. Legal and institutional framework** | **B. System of organisations** | **C. Organisation** | **D. Resources** |
| **Anticipate**<br>1. Risk assessment<br>2. Forecasting<br>**Recognise**<br>3. Monitoring<br>4. Impact assessment<br>**Adapt**<br>5. Prevention & mitigation<br>6. Preparedness<br>7. Response<br>8. Recovery<br>**Learn**<br>9. Evaluation | A.1) Are there any legislation or policy requiring [function]?<br>A.2) Is the utility for [function] stated in legislation or policy?<br>A.3) What stakeholders are identified in legislation or policy as involved in [function]?<br>A.4) Are the legislation or policy stating to whom and how the results of [function] should be disseminated?<br>A.5) Are funds earmarked by legislation or policy for [function]?<br>A.6) Are the legislation or policy implemented?<br>A.7) Are there any values, attitudes, traditions, power situation, beliefs or behaviour influencing [function]? | B.1) What stakeholders and administrative levels are involved in [function]?<br>B.2) Are the responsibilities of stakeholders and administrative levels clearly defined for [function]?<br>B.3) Are interfaces for communication and coordination between stakeholders and administrative levels regarding [function] in place and functioning?<br>B.4) Are interfaces for dissemination, communication, and integration of the output of [function] to stakeholders involved in other functions that depend on the output?<br>B.5) Are interfaces for facilitating coordination between functions in place and functioning? | C.1) What parts of each organisation are involved in [function]?<br>C.2) Are the responsibilities for [function] clearly defined for each involved organisational part?<br>C.3) Are systems for effective collaboration in [function] between the involved organisational parts in place and functioning?<br>C.4) Are there any internal policies for [function] in each involved organisation?<br>C.5) Are these internal policies implemented?<br>C.6) Are interfaces for dissemination, communication, and integration of the output of [function] to parts of the organisation involved in other functions that depend on the output in place and functioning? | D.1) What knowledge and skills on individual level does each involved organisation have for [function]?<br>D.2) What equipment and other material resources does each involved organisation have for [function]?<br>D.3) What funds do each involved organisation has for [function]?<br>D.4) What knowledge, skills and material resources do members of the public have for [function]? |

The scope of this chapter does not allow for presenting the result of the studies per se, but focuses instead on presenting some brief reflections on the utility of the proposed framework. Firstly, feedback from the studies indicates that the use of the framework facilitates increased awareness among the participating stakeholders regarding dependencies and couplings between different functions as well as between different actors, sectors and administrative levels. This is of great importance, especially in a resource-scarce environment, since such awareness may for instance lessen the tendency to work in 'silos', which is not only resource inefficient but also makes it likely to miss important aspects of resilience that are related to the interdependencies in the system. In this sense, the assessments using the concept of societal resilience presented in this chapter worked as a capacity development intervention in itself.

Secondly, the comprehensive assessment of the capacities to perform the nine functions, on the basis of the four levels (A–D) of observable aspects or form, proved to be of great value as it provided important input on how and where to target other capacity development activities to increase societal resilience. It achieved this because it provided the stakeholders with a good understanding of the most important challenges to the system for disaster risk reduction and climate change adaptation in Botswana and Tanzania respectively. To exemplify, in the Botswana case the assessment was followed by a project proposal with activities targeting basically all functions (proactive and reactive) and levels as well as the dependencies. The interventions comprise advocacy activities at the policy level to influence the legislative system, the construction of risk databases and information management systems, targeted trainings and so on. This project is currently running and at the time of writing it is still too early to establish its actual impact on the societal resilience in Botswana. In the Tanzania case, the project proposal is still to be developed at the time of writing.

In summary, the framework for assessing the capacities of the systems for disaster risk management and climate change adaptation described above proved to be of great value in the programming phase of capacity development initiatives to strengthen societal resilience. Its main challenges are however to balance the need for detail in the analysis with the need for grasping the system as a whole, as well as to manage and present the rapidly growing amount of data that is generated when utilising the framework in practice.

## Conclusion

The concept of societal resilience presented in this chapter builds on established theory of Resilience Engineering and operationalises the concept by presenting its purpose, required functions and a way to identify and analyse the complex network of actual forms that together achieve these functions in society. Although this framework for analysing societal resilience has challenges and limitations, the Resilience Engineering approach to societal resilience seems to be both a conceptually and pragmatically fruitful path to follow. The framework itself is also still in the making and more applications of it are on their way and are necessary to develop it further. In short, to meet the rising focus on societal safety and sustainability in a time of increasing complexity and dynamic change in our world, Resilience Engineering approaches to societal resilience constitute a way forward. Societal resilience is in other words an emergent means to reach assurgent ends.

## Commentary

Starting from the key issues of societal safety and sustainability, this first chapter introduced a concept of societal resilience. The concepts of resilience engineering provide a systematic framework that can be used to make sense of concrete cases. Indeed, a consistent frame of reference is the necessary foundation for any attempts to improve or to become better. Understanding how a system works, rather than understanding how it is structured, is necessary for resilience. This line of thinking is echoed by many of the chapters in this book. The following chapter illustrates this by looking specifically at how one can move from descriptions to prescriptions.

# Chapter 2
# Describing and Prescribing for Safe Operations within a Large Technical System (LTS): First Reflections

Jean Christophe Le Coze and Nicolas Herchin

## Introduction

This chapter presents some elements of an ongoing study investigating the way in which safety is produced in normal operation within a 'large technical system' (LTS) operating high pressure gas transmission networks. Although relying strongly on the original manuscript of the 2011 resilience engineering symposium (Le Coze et al., 2011), this new version, apart from updating and expending some sections, opens up more on the 'engineering' part of the study, or what we have described elsewhere for instance as the move from description to prescription (Le Coze et al., 2012). In particular, in this new version, one core issue that is raised, beyond the ability to provide descriptions of resilience (or variability, as it will be next explained), is how outsiders participate in the way an organisation manages safety, for example, through the prescriptions that take shape as a result of the descriptions (and interpretations) that they provided.

This is an important part of any safety research. Indeed, the opportunity to develop safety research with practical purposes (for example, maintaining or improving safe operations) requires being in a position to produce a favourable environment, in which different categories of people (for example, top managers, managers, operators) are, at the start, interested, then convinced about the value of the approach. Although an important part of any safety endeavour, these issues are not always transparent in scientific publications, and this chapter intends to start reflecting on this. This contribution has therefore an experimental nature, mixing both descriptions and how these descriptions, intentionally or intentionally, influence the 'object' of study.

## High Pressure Gas Transmission Network as a Large Technical System (LTS)

The analytical category of LTS was developed about 20 years ago to describe numerous infrastructures (networks) that appeared to share common features. One pioneer in the field is the historian of technology Hughes (1983) who studied the electricity network, from its invention to its wide spread throughout society in North America. This work allowed in the following years a spark of interest in the research community and to gather other researchers already partly involved in this area (two important conferences were held in the 1980s on this subject, leading to the publication of two books, Mayntz, Hughes, 1988, La Porte, 1991). In all these contributions, it is made explicit that beyond the electrical case studied by Hughes, it is interesting to identify a much wider category of systems, which, according to Joerges (1988: 24), '(1) are materially integrated, or "coupled" over large spans of space and time, quite irrespective of their particular cultural, political, economic and corporate make-up, and (2) support or sustain the functioning of very large numbers of other technical systems, whose organisations they thereby link'. For this author, examples of LTS can, as a consequence, be 'integrated transport systems, telecommunication systems, water supply systems, some energy systems, military defence systems, urban integrated public works' and so on.

Of course, with a safety perspective, a very well-studied case of LTS comes to mind: aviation. From pilots' psychological models elaborated by psycho-cognitive scientists to cabin crew communication and coordination issues (leading to crew resource management programs) as well as ATC studies (Air Traffic

Control) by ergonomists (Sperandio, 1977) and high reliability organisation researchers (Roberts, Rochlin, 1987), there is a very large literature dedicated to the safety aspects of this particular LTS. However, other LTS have not been granted as much attention as aviation in the past decades of research on safety, at least to the authors' knowledge. For example, one can think of electrical grids or gas transmission networks, to which serious safety issues are associated.

For instance, a 'black out' can have severe indirect consequences on the exploitation of highly hazardous technological installations depending on such networks. A recent incident in a nuclear power plant in Sweden in 2006 showed very well this issue when its diesel generators failed to start as expected for cooling down the reactor's core, following a 'black out' of the electrical network supplying energy. Although an indirect consequence, the centrality of such an LTS requires that its functioning becomes an integrated part of prevention of major hazard. As far as high pressure gas transmission networks are concerned, they can have direct (as well as indirect) consequences. One such direct consequence is the loss of containment of pressurised gas, leading, potentially, to explosions and fires which may cause several casualties.

## The Threat Of 'External Interference'

One of the specificities of the gas transmission networks in general is that they are opened to potential 'external interference'. Indeed, pipelines are not confined within the boundaries of an industrial site. They cover a very wide geographical area, and represent in France a total of around 40,000 km (Figure 2.1), operated by two different companies (south-west and rest of France).

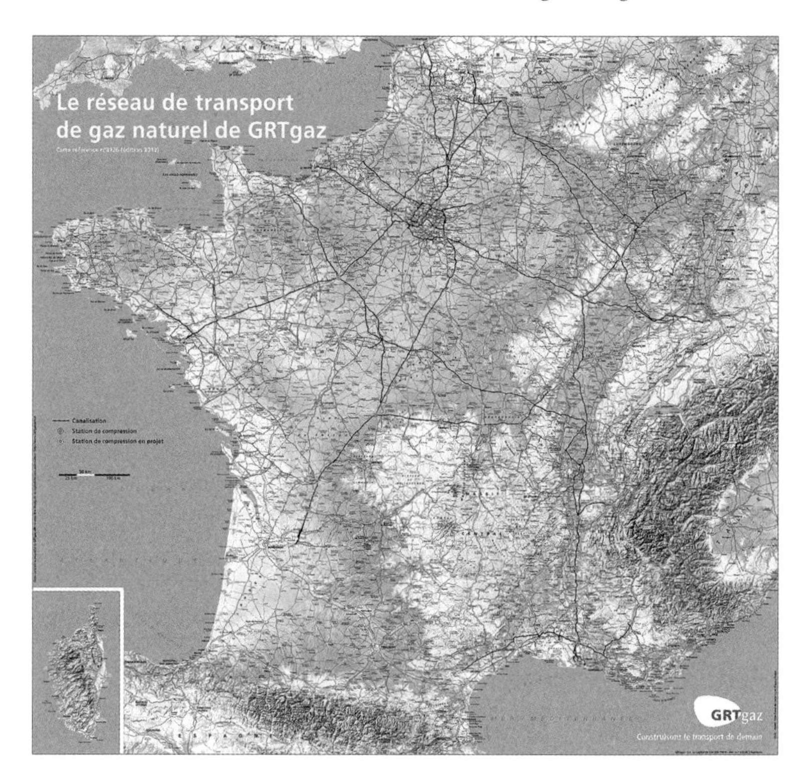


**Figure 2.1 The French gas transmission network**

As a result, one specific safety activity that has been considered in this study is the identification and prevention by the exploiting company of 'external interferences' on gas pipelines. For an LTS such as high pressure gas transmission networks, any urban, municipal or civil engineering (from now on in the text described as 'UMCE') work carried out nearby pipelines and requiring drilling or digging with tractors is a potential threat to their integrity.

*Ghislenghien's Disaster (2004)*

The accident of Ghislenghien in Belgium in 2004 (24 persons killed, 132 severely injured) is one illustration of the type of scenario which can lead to major consequences. This accident was caused by a weakened pipeline's structure following an 'external interference', which would have been caused, from a 'sharp end' point of view and according to the current explanations, by a truck involved in UMCE nearby. After hitting the pipeline without noticing and/or informing the operating company about it, the incident remained totally unknown until the pressure increased in the pipeline and that the weakened structure, where the tractor hit, failed to contain the rising pressure. It created a high pressure gas leak, which led to a rupture and then to a huge flame when ignited by a source of energy (Figure 2.2).

This disaster is an illustration of the challenge faced by companies managing high pressure gas transmission networks for identifying, locating and assessing any works being performed nearby their pipelines.

*Case Study Introduction: Managing the Threat of Mechanical Damage on Gas Transmission Pipeline*

As stated, the threat of external interference, leading to mechanical damage and potentially adverse consequences, is at the heart of safety management systems of gas transmission operating LTS. The present study, involving one of the two French companies operating a 32,000-km gas network, focused on this specific risk of external interference, aiming to grasp the complexity of managing prevention measures.

**Figure 2.2 Ghislenghien's disaster (2004)**

In a nutshell, several levels of preventive measures, ranging from national initiatives to local ones, are put in place to monitor and adequately respond to this risk. Empirical observations have been restricted to local measures in this study. At this level, prevention is based on a decentralised mode of organisation: indeed, for a given territory, a 'sector' (composed of a team of between 6 and 12 individuals, in this case study it was 6) is in charge of the maintenance and the surveillance of pipelines. The sector selected by the company for the study was considered as a 'good' one, with good results and with a good image within the organisation. Located in a complex urban context of a major city, it requires a high level of interaction between many different actors including engineering contracting and subcontracting companies, municipal employees, architects and so on. It is thus representative *a priori* of resilience characteristics of the LTS.

## Observing and Engineering Resilience In LTS

### *Elements of Methodology*

The approach retained for describing, understanding and explaining the activity of a team within a sector (S1) involved in prevention of 'external interference' (through the angle of resilience) relied on participant observations and interviews in 2010. A similar study was then conducted in 2011 in a different type of sector (S2), but methodologies were similar, although refined in the second study given the knowledge of the first empirical case study. In 2012, managerial activities (MA2) were explored (after a first series of series of interviews in 2011 with middle managers interviews MA1) but neither are introduced or discussed in this chapter. In this chapter, we refer mainly to empirical data of S1, although in the engineering section, a broader view is offered of the interaction between the different empirical studies. Interviews were conducted with the manager of the team and then the team members, with a focus on three of them involved, directly or indirectly, in 'external interference' prevention activities.

Interviews were performed at times collectively but also individually, and oriented on specific topics, depending on the function of the interviewee. Questions covered different topics included task complexity, expertise required (technical, relational) for handling work situations, relationship at work between employees and management, training, information flows about work issues and incidents and so on. The same people were sometimes interviewed twice, in order to come back on some aspects missed in the first interview or deepen our comprehension of the activities.

Furthermore, participant observations of activities within and outside the sector premises were performed, consisting in taking into account technological interfaces, communication and coordination between employees, and also understanding the different steps followed for performing tasks while interacting with employees from UMCE companies. When possible, some observed activities were questioned 'on the spot' (as someone learning his job would probably do). For example, following a decision that seemed to rely on a judgment that was unclear from an outsider point of view, it was asked of the employee, when possible, to explain about the rationale behind his/her decision. The idea is to get to know some of these implicit judgments that one observes in work situations and which provides some clues about the acquired expertise of the agents.

The main purpose of these series of interviews and observations was to get close to the specificity of the work activity at the 'sharp end'. When interviews and observations were done separately by researchers, feedback sessions were organised between the observers in order to cross data, and discuss interpretations and hypotheses. Five days were spent observing and interviewing. This 'immersion' within the activity of the sector was prepared few weeks before, by a description and understanding of the more global (regulatory, economical and political) context of this LTS. Several meetings with knowledgeable engineers and managers in this domain allowed the team of observers to get to know the context.

### *Elements of Theory*

Theoretical background for this study includes disciplines such as engineering, psychology, cognitive science, management and sociology applied to safety. The notion of resilience (Hollnagel et al., 2006, Hollnagel, 2009) has been approached in our studies through the concept of variability, and we would tend to see resilience, although this is not definite at this stage, as the positive side of this variability. In theory, and in light of our empirical data, the whole issue seems to have become, from a safety management point of view:

- First, to be able to locate, describe and understand the variability considered to be a problem for safety (negative variability), then try to eliminate or compensate for this variability.
- Second, to be able to locate, describe and understand the positive variability (resilience) in order to maintain, support, share and officialise it when and if found relevant to do so.
- Third, to be able to locate, describe and understand the neutral variabilities, namely the variabilities that are not considered to be a problem but only alternative ways of doing a specific job.
- Fourth, to be able to locate, describe and understand the ambiguous variabilities namely specific practices without clear signs about their positive, negative or neutral sides.

What is interesting with these distinctions from a theoretical point of view is that it is possible, for instance, to indicate variabilities that people agree to be negative but remain engrained in the functioning of the LTS because of the lack of solutions available to make them evolve. These become part of the latencies well documented in safety after disasters, since Reason's formulation (Reason, 1990). There are some examples available in our data for such situations. If these situations are local practices allowing the job to be done, they are not necessarily in theory expected to be maintained. Such a problem is to be dealt at managerial levels.

Of course, there are a lot of obstacles along the way to obtain a clear-cut categorisation in real-life situations that could be applied on a daily basis. First, variabilities observed (whether positive, negative, neutral or ambiguous) only make sense in relation to other variabilities that can somehow compensate each other. Any situation combines multiple variabilities in order to cope with specific contexts, depending also obviously on the choice made in terms of how activities are analysed and decomposed.

Second, because we can only be sure in hindsight about the boundaries created through the combination of positive, negative, neutral or ambiguous variabilities, this type of classification as applied to specific situations in real-life context remains a collective and multidimensional construct, including cognitive, social, cultural and political features. The norms, against which practices (and variabilities) can be scrutinised in real-life situations, whether in hindsight or foresight, are indeed not objectively defined but collectively constructed (Le Coze, 2012).

### *Main Safety Measures for Preventing External Aggressions*

In order to prevent 'external interference' on pipelines, a number of 'in-depth defences' are implemented, ranging from technical to national level:

- Technical: increased resistance of pipelines, mechanical protection, above-ground (yellow) signalisation and so on.
- Regulatory: request for information about pipeline location by companies intending to dig (DR '*demande de renseignements*'), declaration of start of projects (DICT '*déclaration d'intention de commencement de travaux*').
- Operational: answers to requests for information (DR), risk assessment on work sites following DICT by companies digging for UMCE work, surveillance of work sites and pipelines and so on.
- Organisational: training of organisation's employees, information to municipalities and civil, urban or municipal engineering companies about the regulation (DR, DICT), incident analysis at national level and so on.

The case study focused mainly on the operational type of measures, not forgetting the other equally important aspects of pipeline damage prevention, which have yet less been described in a first approach.

## Some Results

### *A Task Requiring Constant Adaptations*

One very explicit feature of the task of preventing 'external interference' appeared to be the ability of some 'home-made' experts to adapt to daily variations of work constraints. Whereas from the description available in the procedures where the task is seen as a sequential series of steps (Figure 2.3), it turns out to be quite different when observed in practice in both S1 and S2.

In theory, indeed, preventing 'external interference' consists in receiving an information request (DR) by a company intending to perform UMCE work on a given area, asking the organisation exploiting the pipelines about the presence or not of a pipeline nearby planned UMCE work. If it is a positive answer, then this company must warn the organisation (through a DICT) at least 10 days prior to the intended start date so that an employee of the LTS exploitation can come and assess the situation on site. The employee of the LTS then locates the pipeline (using specific equipment), and indicates to the UMCE company the safety measure to be taken given the specific situation. The company man then operates supervisory work until the end of the project to identify any damages that would have been caused to the pipeline. All this is represented in Figure 2.3.

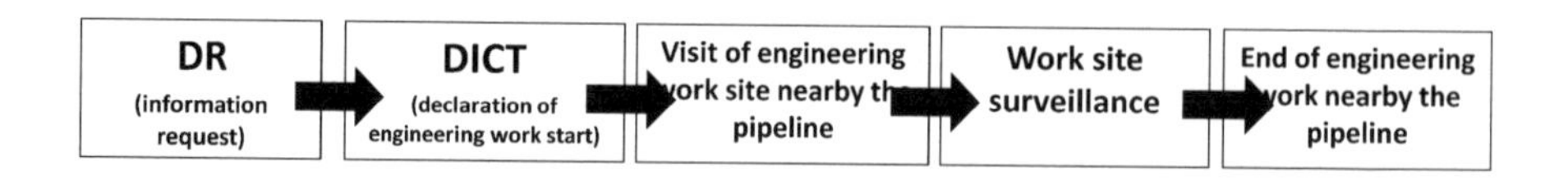


**Figure 2.3 A sequential task, in theory**

This is, however, in theory. In reality, it is difficult to implement all these steps sequentially. First, there are other activities not included, such as responding to invitations from UMCE companies to assist them before projects start, in order to get to know where the pipelines are. But there is also emergency engineering work that need to be treated immediately (such as, for example, water leaks to be fixed by water engineering companies which necessitates sometimes digging close to high pressure gas pipelines). Secondly, companies performing engineering work do not always warn 10 days ahead and it may happen they call the very same day they start the work. As this is not planned, solutions must be found to deal with the situation, and to establish priorities in the initial schedule. Instead of a linear sequence of activities, observations quickly reveal a different adaptive type of task, involving multiple parallel activities (Figure 2.4).

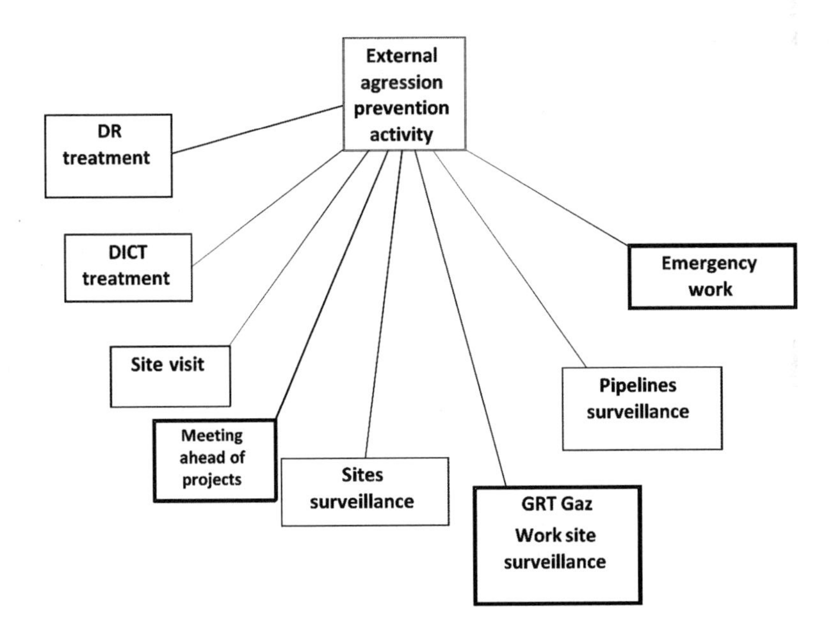


**Figure 2.4 A task involving parallel activities**

*From Collective to Individual Expertise*

All these activities cover a wide range of different aspects and skills: administrative, relational and technical. Depending on the time of year, workload varies. In the most intensive periods, trade-offs must be made between all these activities in order to allocate resources (time, expertise) for what the team considers as the UMCE works nearby to pipelines deserving the closest watch. The coordination between the members of the team is thus very important.

Defined for example by Weick as one dimension of '*collective mindfulness*' (Weick, 2001), this collective side of the activity revealed in the case study interesting features including a mix of adaptations. In this team, it appeared that individual expertise was at the heart of the decision-making process, when circumstances pressed for adaptive patterns. While not in the hands of the manager of the team in S1 (who hasn't got strong experience in the field of 'external interference' prevention), great flexibility was granted to one of the individuals in the team who was acknowledged by others as the expert in this area.

Organising his/her own schedule, s/he is in charge of balancing several factors for deciding in real time about the priorities when trade-offs must be made. Of course, s/he relies on written procedures provided by the organisation but, given the specificity of local urban contexts, s/he has to adapt this procedure in order to achieve what seems to him/her to be a satisfying response (Simon, 1947) to his/her local constraints and unplanned demands. His/her choices have an impact on the team workload, as s/he can, at times, when it is required, ask a colleague (or even his/her manager) to replace him/her for a planned visit. While s/he is replaced, s/he is then able to deal with an unexpected situation, such as emergency engineering work in a sensitive area for which s/he wishes to be present to ensure close supervision.

This employee commented this situation as an unusual one, as s/he could 'give order to his own boss'. Whereas probably in many cases one could imagine the problem of such a situation, in this team, it was not an issue. A balance had been found between hierarchy and expertise, introducing here a key issue of power in teams. This is one side of the collective expertise of the team, which partly results from the managerial style of the team leader, willing to have an expert in the team leading some decisions for the rest of team, although not in a managing position.

Beyond this collective side of expertise, and as a complement, what has proved also very interesting to identify and to investigate are elements of decision-making processes on which this individual's expertise relies. Without attempting to produce a final model of such a complex topic as decision-making, the strategy in this study was rather to see how data collected during interviews and observations could be exploited in order to indicate some of the key dimensions of this individual expertise.

In this respect, the (methodological and theoretical) approach was close to principles of 'macro cognition' or 'cognition in the wild', considering cognition to be understood as individuals ('cleverly') adapting within their work context. Rather than expecting to fully understand cognition only through experiments (and also normative frameworks), field studies ('in the wild') become one input for understanding how cognition proceeds in the face of complex and dynamic environments (including interactions with various types of media – procedures, maps, screens) instead of simple and static ones.

### *Towards a Heuristic Model of Activity in the Domain of 'External Interference' Explicitly Indicating Variabilities*

One first step was to group together many of the data that this expert processed in his/her activity. A first set of six groups of key dimensions were identified. They combine many different sides implied by 'external interference' prevention, among which are 'itinerary', 'urban geography', 'pipeline', 'urban features', 'political context' and 'engineering'. Each group refers to specific topics, such as rush hour, depth of pipelines, UMCE companies, municipalities' services and so on (Figure 2.5).

It is clear from this description that only individuals with years of experience are in a position to provide comfort for the decision-making process involved. One reason for having a specialised 'expert' in this team and to allow him/her flexibility of decision is to find an answer to this problem of knowing sufficiently well the area, the UMCE companies (see groups in Figure 2.5) and so on in order to elaborate appropriate choices in real time when needed.

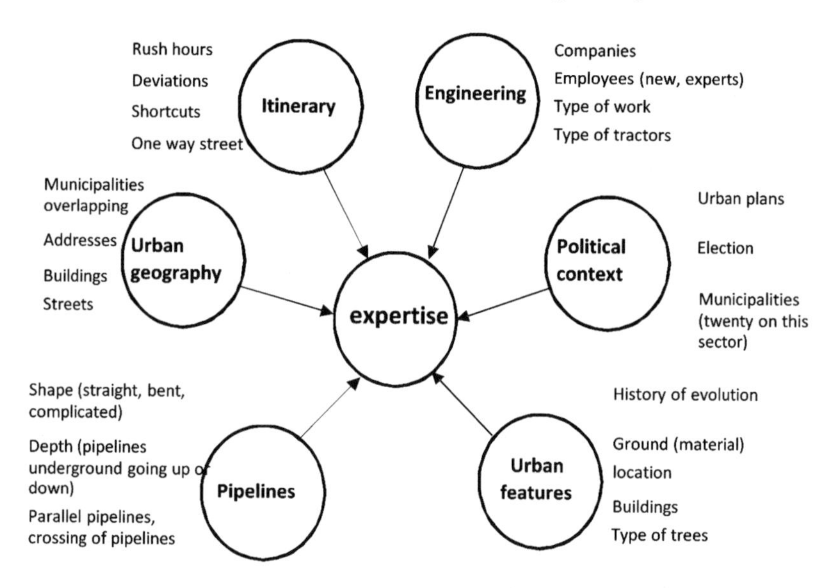


**Figure 2.5 Expertise at the centre of many different topics**

A similar study conducted in 2011 in a different type of sector (S2) with the same methodology allowed us to draw on the first conclusions to further deepen our understanding of these cognitive processes. Although the second sector was very different in terms of environment and constraints, strong similarities could be identified in terms of expertise. In particular, this expertise was found to rely upon two important cornerstones:

- A great prudence towards uncertainties facing 'external interference' activities (for example, on UMCE companies themselves, their behaviour, on possible accident scenarios, on the evolutions of the activity or even on judicial exposition in case of problems).
- A great sensitivity towards interpersonal interaction strategies to be put in place, and the development of key expertise in this field (for example, developing strategies to collect information, convince, reprimand and so on as, for instance: questioning practices, finding the right way of speaking, increasing dramatic intensity in the speech and so on).

At first sight, many aspects of the expertise described in the first sector can thus be seen, thanks to additional data from S2, as the (collective) ability to adapt one's response in-between acceptable limits of 'variability' (including the many types of variabilities, introduced earlier in section 4.2). One way to suggest a first model of the activity was as a consequence to include the notion of expertise at the heart of it, based on a combination of collective and individual dimensions. For describing expertise, Klein's approach proved useful (Klein, 2009). Synthethised through four key features: mental simulations, intuition, mastery of time horizon and knowledge by expert of his own limit, they were illustrated by some of the stories collected during observations and interviews.

For example, cases of 'intuition' or 'mental simulations' were used as illustration of how expertise shaped decision-making. Combining the various features of the activity together led to the following model Based on a wide range of data collected by the expert on many different topics (Figure 2.6), divided in categories of 'global context' and 'specific context', individual expertise combined with collective expertise allow the team to adjust in real time their schedule in order to maximise chances of being at 'the right place at the right time'. Figure 2.6 shows an updated model of the activity of 'external interference' highlighting this central concept of 'variability'.

This notion is at the heart of this LTS's issues. Indeed, due to historical and structural aspects, it has the strong particularity of being both a decentralised system and a very open system (for example, UMCE). In such a configuration, the issue of variability introduced here plays a central role, perhaps more than in other systems, indeed:

- Operators are left to decide for themselves how they organise their activity given specific contexts: there are more than 80 sectors, for which direct supervision by management is impossible.
- Working situations cannot be entirely controlled by the company, which has no direct grasp on its external environment (for example, UMCE).

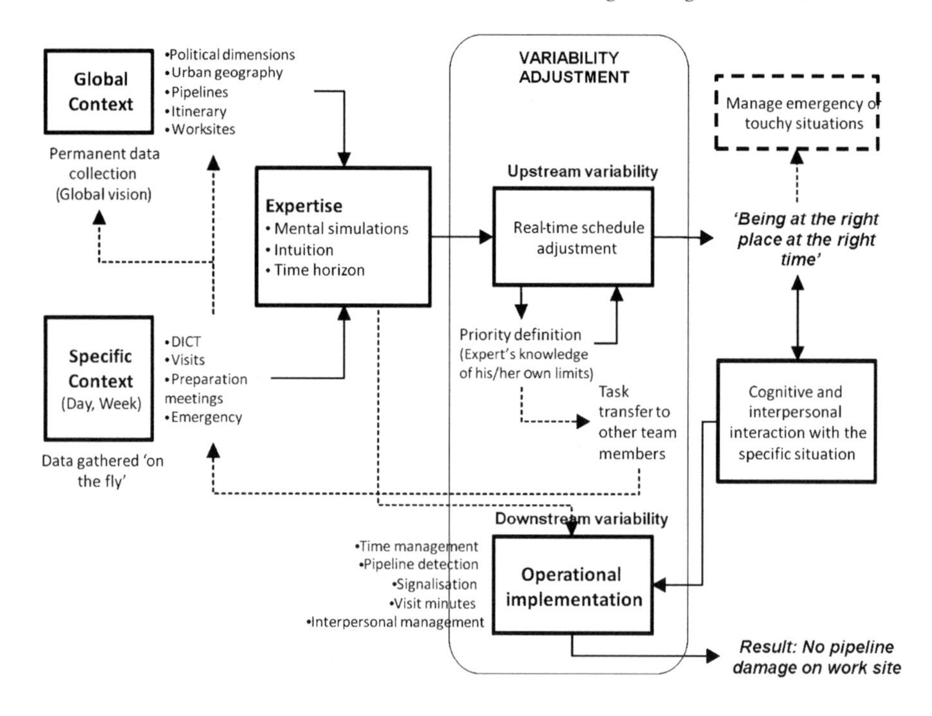


**Figure 2.6 An updated model of sector activity putting forward the central concept of 'variability' emerging from empirical studies**

In short, variability can be seen through a number of adaptations in operational practices, as a consequence of the level of autonomy left by management to operators to perform their work. In practice, limits to this variability are however explicitly or implicitly defined through:

- Acquired knowledge in sectors about how to adapt to local and historical specificities;
- A certain level of rationalisation of task sought by central management in order for the variability caused by necessary local initiatives to remain in-between acceptable limits as far as safety is concerned.

This notion of acceptable variability boundaries links back to the level of centralisation or decentralisation, as well as the question of regulation modes thus raising organisational issues.

## From Description to Prescription

In this last section, we want to discuss some of the points made earlier in the introduction about the 'engineering' part of this study, that we translate in the question of how some descriptions (and interpretations) provided at one stage are translated into prescriptions, namely official changes on the part of the organisation (that is, networks of actors, inside or outside the organisation, which have the power to make these changes) regarding its approach of the management of safety. At this stage, the section will deliberately remain sketchy, but is meant to indicate future directions for more elaborated developments based on more extensive presentations of empirical results and of our context (including the complex relationships between the company investigated and the members of this project).

It is also meant to provide a better grasp and refinement over the link between our ability to describe and make sense of real-life situations thanks to the help of relevant models, and the ability to shape activities in a direction that we could consider improving operations from the point of view of safety. This includes several issues:

- Designing a research framework in which researchers and practitioners will interact in an appropriate manner in order to collectively move towards, if possible, safer operations.
- Finding a balance between the need for describing and interpreting real-life situations, and the development of useful tools or instruments (whether conceptual or methodological) that actors of the company (and sometimes also outside the boundaries of the formal company – for example, in this case, the UMCE actors) will use in their daily operations.
- Remaining sensitive to the developing needs and ideas produced by different categories of actors in the company regarding their own practices while interacting with them through interviews, observations and feedback sessions. Tools have to be appropriate to these different categories of actors and real life contexts.

In this respect, if we stand back, we can represent our different studies in the past three years (2010–12) as follows:

2010

- Empirical fieldwork (interviews and observations) in operations (first sector, S1)
- Interpretations of data
- Feedback to practitioners met during observations and interviews
- Report with interpretations
- Feedback to the internal department which funded the study,
- Definition of further needs.

2011

- Second empirical fieldwork (interviews and observations) in different type of operations (second sector, S2)
- Interviews with middle managers of Entity (MA1)
- Feedback (at different levels)
- Report and proposition of practical recommendations
- Design of a training course on Human Factor for operations.

2012

- Third empirical fieldwork (interviews) with top managers of an Entity (managerial activities MA2)
- Interpretations of data
- Working group on 'variability' (WG1)
- Feedback (at several different levels)
- Report and proposition of practical recommendations.

It turns out that these stages have regularly consisted in different cycles, alternating:

- descriptions of work situations
- interpretation of data
- feedback sessions with different actors
- opportunities for further developments discussed with different actors of the organisation, but also,
- translations of discussions into relevant programs for the future

- and, at times, new type of prescriptions.

This cyclical structure repeated several times on different topics has been incremental and has now led to some new prescriptions. One is briefly commented here.

Based on our descriptions of the expertise of operators, the company has shifted its strategy regarding procedural content definition and communication. Instead of prescribing precisely what is expected from agents to the extent that it is impossible to comply with the document in most real-life situations, much more space is now left to the judgments of agents that are in a position to appreciate what type of tradeoffs need to be made more locally. But this move was made possible because our empirical descriptions made visible what was left invisible, namely the professionalism and expertise of the category of operators dealing with external interferences. In this respect, interpretations had a performative effect by opening new prospect in the prescriptive philosophy of the company in relation to operations. Of course, this new approach has to be backed up by the ability to create spaces where the various types of variabilities can be regularly discussed and addressed. Spaces that now also need to be designed and experimented (a pilot study was studied in 2012, WG1).

## Conclusion

This study presented here has the purpose of observing some of the features of operations and to imagine ways of being practical in order to promote, maintain or create safe activities through new concepts, including the recently framed concept of resilience. After describing the specificity of LTS, and associating high pressure gas transmission networks to this category, the chapter introduces the threat of 'external interference' as one activity to be studied. LTS are opened systems that need to be managed in order to cope with this specific threat. The study relies on interviews and observations for approaching what could be seen as the properties of individuals and team dealing with the prevention of 'external interference'. The chapter shows a collective and an individual expertise, allowing flexibility and quality of decision-making to be obtained when necessary. A heuristic model, as a first attempt to capture characteristic of activity within this LTS, is suggested. It includes the issue of variability, in relation to the topic of resilience. The elements of descriptions obtained through empirical fieldwork is then introduced in the broader issue of a better understanding of how we move from description to prescription. A cycle, or pattern, combining different steps is identified and shortly commented on in relation to effective prescriptions which derived from the project.

## Commentary

The step from description to prescription brings forward the potential conflict between what people actually do and what the prescriptions require them to do. This is also known as the discrepancy between work-as-done and work-as-imagined. While it is essential that some prescriptions for work are created in order to further the purposes of productivity and safety, it is equally essential that such prescriptions recognise the benefits of performance variability and performance adjustments. People at the sharp end have to continuously organise their work to overcome various forms of external interference. The prescriptions of work should help that rather than hinder that.

# Chapter 3
# Fundamental on Situational Surprise: a Case Study with Implications for Resilience

Robert L. Wears and L. Kendall Webb

**Things that never happened before happen all the time (Sagan, 1993).**

## Introduction

Surprise is inherently challenging to any activity; it challenges resilient action, since by definition it cannot be anticipated, and for some types of surprises, monitoring is limited by both lack of knowledge about what to target and the absence of precursor events or organisational drift (Dekker, 2011; Snook, 2000) that might have provided even soft signals of future problems. It does, however, present opportunities both for responding and for learning. In this chapter we describe a critical incident involving information technology (IT) in a care delivery organisation. The incident was characterised by the co-occurrence of both situational and fundamental surprise (Lanir, 1986), and the responses to it are informative about both specific vulnerabilities and general adaptive capacities of the organisation. We studied this event to gain insight into three aspects of resilience: first, how adaptive capacity is used to meet challenges; second, to understand better what barriers to learning are active; and finally, to infer recommendations for practice. We conducted these analyses both shortly after the event, and then revisited them in discussions with key participants about three years later. We note that temporal and cross-level factors played important roles in affecting the balance between situational and fundamental learning. Because the situational story (of component failure) developed first, it was difficult for the fundamental story of unknown, hidden hazards to supplant it. In addition, the story of the situational surprise was easily understood by all members of the organisation, but that of the fundamental surprise were difficult for many to grasp, including (especially) senior leadership, who tended to adopt an over-simplified (situational) view of the problem. Finally, over time, the fundamental surprise was virtually forgotten, and those members of the organisation who do remember it have (in effect) gone into self-imposed exile. Thus, although the organisation did learn and adapt effectively from this event, it has become progressively blind to the continuing threat of fundamental surprise in complex technology.

*The Nature of Surprise*

Analyses of critical incidents often distinguish between situational and fundamental surprise (Lanir, 1986; Woods, Dekker, Cook, Johannesen and Sarter, 2010). Events characteristic of situational surprise might be temporally unexpected, and their specific instantiation might not be known in advance, but their occurrence and evolution are generally explicable and, more importantly, compatible with the ideas generally held by actors in the system about how things do or do not work, and the hazards that they face in ordinary operations. For example, a sudden rain shower on a previously sunny day would be surprising, but generally compatible with our experience of weather. Fundamental surprise, on the other hand, is astonishing, often inexplicable, and forces the abandonment of the broadly held notions of both how things work, and the nature of hazards that are confronted. For example, a volcanic eruption in Paris would challenge basic notions about the geology of volcanism because it is incompatible with prior understandings.

An apocryphal story tells of a famous lexicographer (Noah Webster in America; Samuel Johnson in Britain) being unexpectedly discovered by his wife while locked in a passionate embrace with a housemaid. His wife exclaimed, 'Oh! I am surprised!' To which he reportedly replied, 'No, my dear. I am surprised; you are astonished'.

It should be noted that the situational *vs* fundamental typology is a relative, not a dichotomous distinction. In addition, the same event for some people might be a situational surprise but for others a fundamental surprise, depending on the relation to and experience with the domain.

If we think of a system's resilience as its intrinsic ability 'to adjust its functioning prior to, during, or following changes or disturbances, so that it can sustain required operations under both expected and unexpected conditions' (Hollnagel, 2011), then it is clear that surprise creates unexpected demands that call for a resilient response.

Lanir (1986) has identified four characteristics that distinguish situational from fundamental surprise. Fundamental surprise refutes basic beliefs about 'how things work', while situational surprise is compatible with previous beliefs. Second, in fundamental surprise one cannot define in advance the issues for which one must be alert. Third, situational and fundamental surprise differ in the value brought by information about the future. Situational surprise can be averted or mitigated by such foresight, while advance information on fundamental surprise actually causes the surprise. (In the preceding examples, advance knowledge that it was going to rain would eliminate the situational surprise, and allow mitigation by carrying an umbrella; advance knowledge that a volcano was going to erupt tomorrow in the *Jardin du Luxembourg* would, in itself, be astonishing, as astonishing as the actual eruption.) And finally, learning from situational surprise seems easy, but learning from fundamental surprise is difficult.

## *Resilience and Surprise*

Resilience is characterised by four essential capabilities: monitoring, anticipating, responding and learning. While effective management of situational surprise would typically involve all four of these activities, fundamental surprise clearly is a profound challenge for resilience, because one cannot monitor or anticipate items or events that are inconceivable before the fact. Even though one can still monitor the system itself, rather than events, the lack of precursors, leading signals or drift in fundamental surprise severely hampers this modality. This leaves only responding and learning as the immediately available resilience activities in fundamental surprise,[1] and explains in part why fundamental surprise is such a challenge to organisational performance. However, fundamental surprise does afford opportunities for deep learning, in particular the development of 'requisite imagination', an ability to picture the sorts of unexampled events that might befall (Adamski and Westrum, 2003)

We present a case study of:

- The catastrophic failure of an information technology system in a healthcare delivery organisation;
- The organisation's response to it from the point of view of resilience;
- And the organisation's memory of the event years later.

The failure itself involved a combination of both situational and fundamental surprise. As might be expected, the immediate response involved both adaptations of exploitation (that is, consuming buffers and margin for manoeuvre to maintain essential operations) and adaptations of exploration (that is, novel and radical reorganisations of the way work gets done) (Dekker, 2011; March, 1991; Maruyama, 1963). Because fundamental surprise makes the disconnect between self-perception and reality undeniable, it affords the opportunity for a thorough-going reconstruction of views and assumptions about how things work, as effortful and unpleasant as that generally seems. However, in this case the conflation of fundamental and situational surprise led to a classic *fundamental surprise response* – a reinterpretation of the problem in local and technical terms, which allowed an easy escape from the rigours of fundamental learning.

1 In the strictest sense, a limited variety of anticipation might still be possible, in Rochlin's sense of the 'continuing expectation of future surprise' (Rochlin, 1999), or 'expecting the unexpected'.

## The Case

In this section we describe the events and the adaptations to the interpretations made of them, based on notes, formal reviews and interviews during and after the incident.

### *Events*

Shortly before midnight on a Monday evening, a large urban academic medical centre suffered a major information technology (IT) system crash which disabled virtually all IT functionality for the entire campus and its regional outpatient clinics (Wears, 2010). The outage persisted for 67 hours, and forced the cancellation of all elective procedures on Wednesday and Thursday, and diversion of ambulance traffic to other hospitals (52 major procedures and numerous minor procedures were cancelled; at least 70 incoming ambulance cases were diverted to other hospitals). There were 4 to 6 hour delays in both ordering and obtaining laboratory and radiology studies, which severely impacted clinical work. The total direct cost (not including lost revenue from cancelled cases or diverted patients) was estimated at close to $4 million. As far as is known, no patients were injured and no previously stored data were lost.

The triggering event was a hardware failure in a network component. This interacted with the unrealised presence of software modules left behind from an incompletely aborted (and ironically named) 'high availability computing' project some years previous; this interaction prevented the system from restarting once the network component was replaced. The restart failure could not be corrected initially because of a second, independent hardware failure in an exception processor. Once this was identified and replaced, the system still could not be restarted because unbeknownst to the IT staff, the permissions controlling the start-up files and scripts had been changed during the same project, so that no one in IT was able to correct them and thus restart the system. This fault had gone undetected because the system had not been subjected to a complete restart (a 'cold boot') for several years.

### *Adaptations*

After a brief initial delay, the hospital was able to quickly reorganise in multiple ways to keep essential services operating for the duration. Adaptations included exploitation of existing resources or buffers; and exploration of novel, untried ways of working. These adaptations correspond roughly to the first- and second-order resilient responses described by a well-known materials science analogue (Wears and Morrison, 2013; Woods and Wreathall, 2008).

Adaptations of exploitation included deferring elective procedures and speeding discharges of appropriately improving inpatients. The former was limited in scope because the extent of the problem was not realised until Tuesday's elective cases were well under way. The latter was stymied by the slow delivery of laboratory and imaging results; physicians were reluctant to discharge patients whose results were still pending. This, of course, is one of the classic patterns of failure – falling behind the tempo of operations (Woods and Branlat, 2011).

Several adaptations of exploration were invoked. An incident command team was formed. Because the area experiences frequent hurricanes, the incident command system was well rehearsed and familiar, so it was adapted to manage a different type of threat.

A similar novel use of available techniques evolved dynamically to compensate for the loss of medical record numbers (MRNs) to track patients, orders and results while the system was down. The emergency department had been planning to implement a 'quick registration' method, in which only basic patient information is obtained initially to permit earlier orders and treatment, and the remainder of registration completed at a later time. The IT failure prevented complete registration but was thought to have left the capability for quick registration. The incident occurred very close to the previously scheduled 'quick registration' implementation, so it was pressed into service early. However, its application in this setting uncovered a problem, in that different organisational units used the same variable to represent different information; this resulted in several patients getting 'lost' in the system. This failure led to an alternative, the use of the mass casualty incident (MCI) system.

In many MCIs, the numbers of arriving patients rapidly exceed the ability to record even their basic information and assign them identifying MRNs, so the organisation maintained a separate system with reserved MCI-MRNs and pre-printed armbands. Although this system was envisioned for use in high demand situations, in theory it could accommodate any mismatch between demand and available resources. In this case, demand was normal to low, but resources were much lower, so the MCI system was used to identify and track patients and marry them to formal MRNs after the incident had been resolved.

The most novel adaptation of exploration included rescheduling financial staff (who now had nothing to do, since no bills could be produced or charges recorded) as runners to move orders, materials and results around the organisation that had previously been transmitted electronically.

### *Interpretations*

The case was viewed in multiple ways within the organisation, depending on the orientation to situational or fundamental surprise. It should be emphasised that there is not a 'correct' interpretation here – these views have both validity and utility, and must be understood and held simultaneously for a full understanding of the case and its implications for organisational resilience.

#### *Situational surprise*

Because the triggering event was a hardware failure, and because the organisation had experienced a similar incident leading to total IT failure 13 years previously (Wears, Cook and Perry, 2006), the failure was initially interpreted as a situational surprise. It evinced no fundamental misperception of the world; it was not 'the tip of the iceberg' but rather a hazard about whose possibility there had always been some awareness.

However, we should not downplay the importance of the organisation's situational response, which was in many ways remarkably good. The organisation detected the fault and *responded* relatively quickly and effectively; the unfolding understanding of the situation and effectiveness of the response was *monitored*, and the organisation reconfigured to meet the threat. This reconfiguration involved a mixed control architecture where a central, incident command group set overall goals and made global-level decisions (for example, cancelling elective procedures, reassigning financial staff) and managed communications among the various subunits of the organisation, while allowing functional units (for example, the emergency department, operating room, intensive care units, pharmacy, radiology, laboratory and nursing) to employ a mixture of pre-planned and spontaneously developed adaptations to maintain performance.

There was a specific attempt to capture situational *learning* from the incident. Each major unit conducted its own after-action review to identify performance issues; the incident command group then assembled those and conducted a final, overall review to consolidate the lessons learned. This review obtained broad participation; it resulted in 104 unique items that, while locally oriented and technically specific, form the nidus of organisational memory and could inform the approach to similar future events, which are broadly *anticipated* in their consequences (that is, another widespread IT failure at some point seems assured) if not in their causes.

One remarkable aspect of the response was the general absence of finger-pointing or accusatory behaviours, witch-hunts or sacrificial firings. An essay on how complex systems fail (Cook, 2010) had been circulated among the senior leaders and the incident command group during the outage, with substantial agreement on how well it described the incident, its origins and consequences; this essay played an important role in minimising the temptation to seek culpability (Dekker, Nyce and Myers, 2012).

*Fundamental surprise*

However, as a fuller understanding of the incident developed, situational gave way to fundamental surprise. The discovery of the permissions problem refuted taken-for-granted beliefs – that the IT section understood and could maintain its own systems;

and in particular, that restrictions to privileged ('root') access could not be compromised except by sabotage. It raised the question of what other, previously unknown threats, installed by a parade of vendors and consultants over the years, lay lurking just beneath the surface waiting to be triggered into behaviours both unexpected and unexplainable.

Lanir notes that 'when fundamental surprises emerge through situational ones, the relation between the two is similar to that between peeled plaster and the exposed cracks in the wall. The plaster that fell enables us to see the cracks, although it does not explain their creation' (Lanir, 1986). The IT unit recognised this clearly, and were astonished by the 'hidden time bomb' whose presence was only fortuitously revealed by the line card failure. This triggered a deeper review of known previous changes, a new commitment to not permitting unmonitored and undocumented changes by vendors or other third parties, and more stringent requirements for 'as installed' documentation (including personal identification of involved parties). It led to a general awareness among IT leaders that their knowledge of their own system was incomplete and that they should therefore act in the 'continuing expectation of future surprise' (Rochlin, 1999) or as in Sagan's remark at the beginning of the chapter that 'things that never happened before happen all the time' (Sagan, 1993). This fundamental learning, however, did not spread throughout the organisation, but remained mostly encapsulated in IT.

*The long view*

In the years following the event, key personnel involved in the recovery made career moves that may have been influenced in part by this experience. The IT disaster recovery specialist was shaken by the fundamental surprise of this event, and oscillated between responding with human sacrifice – 'falling on her sword' – and resentment that her prior warnings and recommendations had been incompletely heeded. Eventually, she voluntarily left the disaster recovery post to take a position in the implementation group. The then IT director, whose leadership in the crisis was little short of extraordinary, decided to leave the organisation for a more technical and less managerial position in another industry. Neither had been subjected to discipline, threats or recrimination by the organisation, but initiated these changes on their own. Unfortunately, their departure left a void in organisational memory, such that situational surprise became the dominant view of the incident as time passed.

Some beneficial organisational learning did persist. The incident command centre and poly-centric control architecture employed in the management of this event was widely viewed as successful, and so was reused on several occasions subsequently, each time successfully. These occasions included anticipatory use in advance of major system upgrades. Thus the organisation can be said to have learned better how to respond, to have improved its repertoire of possible responses and its sensitivity to anticipating potentially problematic events. However, the experience of successfully anticipating and managing events over a long period of time risks the development of overconfidence, especially when the impact of fundamental surprise has been forgotten.

## Discussion

Critical incidents are ambiguous: managing an event that stops just short of complete breakdown is both a story of success and a harbinger of future failure (Woods and Cook, 2006). Incidents embody a dialectic between resilient adaptation and brittle breakdown. In this case we see successful, resilient adaptation, but the real lesson is not in the success (Wears, Fairbanks and Perry, 2012) but rather in how adaptive capacity was used, how it can be fostered and maintained and how learning occurs. We also see limited fundamental learning, but the real lesson is not the failure of more broadly based learning but rather understanding what made that learning difficult.

### *Fundamental Surprise as a Challenge to Resilience*

Fundamental surprise represents a major challenge to resilient performance. Since by definition, fundamental surprise events are inconceivable before the fact, they cannot be anticipated; since it is unknown whence they come, there can be little guidance on what, exactly, to monitor to facilitate their detection.

### *Factors Limiting Fundamental Learning*

There is a strong tendency to reinterpret fundamental surprise in situational terms (Lanir, 1986). Several factors combined to limit fundamental learning in this case.

*Situational surprise*
The co-occurrence of a situational surprise (failure secondary to component failure) made it easy to redefine the issues in terms of local technical problems (for example, the lack of available spares). The easy availability of hardware failure as an explanation for the outage limited deeper analysis and understanding. This is likely an expression of an efficiency-thoroughness trade-off (Hollnagel, 2009; Marais and Saleh, 2008); accepting a simple, understandable explanation saves the resources that would be used in developing a deeper, more thorough understanding. In addition, the relative success of the adaptations to the failure paradoxically made deeper understanding seem less important.

*Temporal factors*
The full understanding of the incident did not develop until roughly 36 hours into the outage, so the initial characterisation of the problem as a hardware issue proved hard to dispel. In addition, the 24 × 7 × 365 nature of healthcare operations required urgent responses to prevent immediate harm to patients. This narrowed the focus of attention to actions that could be taken immediately to manage the disturbance, and moved deeper understanding to a lower priority. Because of this narrowed focus, the major formal opportunity for learning, the after-action review, was limited almost entirely to issues related to the adequacy of the response; little effort outside of the IT section was invested in understanding the causes of the failure, and even less on understanding what the incident revealed about other hidden threats, threats that may not yet have been activated.

*Cross-level interactions*
Different understandings were held at different levels of the organisation. The technical problem – unauthorised, unrecognised access to critical files – was harder for non-technical leadership to

understand, particularly compared to the easily grasped story of component failure. Although one might suspect that the full story might have been embarrassing thus obscured or suppressed, this was not the case; the IT leadership was remarkably forthcoming in laying out the full explanation of what was known, as it became known.

In addition, one might question whether it was even pertinent for the clinical arm of the organisation to undergo fundamental learning. Clinical operational units need to be prepared for the consequences of IT failures, but have little role in anticipating or preventing them.

*Healthcare-specific factors*

IT in healthcare has several unique characteristics that contributed to both the incident and to the difficulty of fundamental learning. In contrast to other hazardous activities, IT in health is subject to no safety oversight whatsoever. The principles of safety-critical computing are virtually unmentioned in a large medical informatics literature (Wears and Leveson, 2008). Thus there is no locus in the organisation responsible for the safety of IT, and no individual or group who might be responsible for deeper learning from the incident.

In addition, IT in healthcare is relatively new compared to other industries. The systems in use today are fundamentally 'accidental systems', built for one purpose (billing), and grown by accretion to support other functions for which they were never properly designed. This has led to 'criticality creep', in which functions originally thought to be optional gradually come to be used in mission-critical contexts, in which properties that were benign in their original setting have now become hazardous (Jackson, Thomas and Millett, 2007).

*Diverting factors*

Finally, an external factor diverted at least senior leadership's attention from a deeper exploration of the vulnerabilities whose presence this incident suggested. Nine months prior to this incident, the larger system of which this hospital is a part made a commitment to install a monolithic, electronic medical records, order entry and results-reporting system, provided by a different vendor across the entire system. Although full implementation was planned over a five-year span, major components of the new system were scheduled to go live nine months after the incident. This project gave the (misleading) appearance of a clean replacement of the previous system, *a deus ex machina*, and thus limited the felt need to understand the vagaries of the existing system more deeply, in addition to consuming a great deal of discretionary energy and resources.

*Implications for Practice*

Fundamental surprise is fortunately a rare event; its infrequency not only makes learning more difficult but also more important. An important general principle we glean from these events is the advantages of 'experiencing history richly' (March, Sproull and Tamuz, 1991) by attending to more aspects of an event. This requires a broader focus for causal investigations; rather than narrowly coning in on the cause(s) of a specific failure (which causes are unlikely to reoccur in this same configuration again), the investigation should broaden its scope, using the incident to identify broad classes of risks to which the organisation is exposed. Similarly, the enumeration of specific errors, failures and faults leading to an event does nothing to illuminate the processes that produce those errors, failures and faults, or permit their continued existence (Dekker, 2011).

Another way to enrich the learning from fundamental failure is to encourage multiple interpretations and accounts from multiple points of view. People tend to see only those issues they feel capable of managing, so by engaging a variety of disciplines and backgrounds, a team can see more (because they can do more). This runs counter to the felt desire to come up with an agreed consensus account of events, and runs the risk of local agreement within subunits of a complex organisation but global disagreement on hazards; thus, some mechanism for both sustaining a variety of accounts and sharing them broadly across the organisation would be important.

Organisations might also foster the construction of 'near histories' or hypothetical scenarios that might have evolved out of the incident in question; this would help develop the capacity for imagination that could help sustain the continuing expectation of surprise and counter overconfidence.

## Conclusion

Fundamental surprise is a challenge for organisational resilience because anticipation is not a factor (or is at least severely restricted) and monitoring is limited, typically, to evaluating the quality of response. Fundamental surprise also affords great opportunities for deep and fundamental learning, but it is difficult to effectively engage organisations fully in the learning process. In this case, the combination of situation and fundamental surprise blurred the distinction between them; situational adaptation and learning were remarkable, but the ease of reinterpreting fundamental as situational surprise meant fundamental learning was encapsulated, limited to only parts of the organisation and subject to gradual attrition through the loss of key personnel.

## Commentary

Every now and then (but hopefully rarely) a system may encounter situations that are completely surprising, and challenge pre-conceived ideas about what may happen and what should be done. For such fundamental surprises, resilience depends more on the abilities to respond, monitor, learn and anticipate internal rather than external developments. This can also be seen as the borderline between traditional safety management and disaster management. While fundamental surprises are challenging, they also create unique possibilities for strengthening all four basic abilities. The boundary between accidents and disasters is breached in the following chapter, which analyses the Fukushima Daichi disaster.

# Chapter 4
# Resilience Engineering for Safety of Nuclear Power Plant with Accountability

Masaharu Kitamura

Various accident investigations have been conducted to identify the causes of the Fukushima nuclear accident and to propose countermeasures to prevent future severe accidents. This chapter describes an attempt to review the investigations from the perspective of Resilience Engineering. Most investigations are based on a fundamental approach of listing up adverse events experienced during an accident to find out the causes of each adverse event, and to propose countermeasures to eliminate the identified causes. An implicit belief underlying the fundamental approach is that safety can be achieved by eliminating the causes that contributed to the accident. As a natural consequence, the causal descriptions and proposed recommendations are large in number and complicated in structure. It would obviously be desirable if the proposed recommendations were better organized and simplified. The present study took an alternate approach where safety of a system is believed to be achievable by properly maintaining the four essential capabilities proposed in Resilience Engineering: responding, monitoring, anticipating and learning. Also, efforts have been made to find so-called "second stories" (Woods and Cook, 2002) latently existing behind the multiple causal relationships described in the investigation reports.

## Introduction

### *Description of Problem*

The nuclear accident which occurred on March 11, 2011, at the Fukushima-Daiichi Nuclear Power Station (NPS) of Tokyo Electric Power Company (TEPCO) was truly an unprecedented disaster. As of July 2013, about 160,000 residents in Fukushima Prefecture have still not been able to return to their hometowns. Citizens' concern about the nuclear accident remains high two years after the accident. Most of the NPSs in Japan have been shut down without any clear timetable for restarting operations. Japanese nuclear experts are obliged to clarify factors that contributed to the disaster. They are also obliged to propose dependable countermeasures to prevent recurrence of another disaster. Through these activities, they have to recover public trust on nuclear industry. Although these are extremely difficult tasks, they should be done irrespective of future nuclear policy in Japan. Without recovering the public trust, no nuclear policy, including nuclear phase-out, will ever function properly.

Various accident investigation reports have been published. A large number of statements and opinions concerning causes of the accident have been put forth and recommendations have been proposed to eliminate the causal factors and thus to improve the safety of NPSs in the future. However, the causal analyses are in an interim state, as are the recommendations. The usefulness and sufficiency of the recommendations to prevent future large-scale disasters have not yet been confirmed. Further efforts are necessary to obtain more systematic and accountable recommendations that can be understood by citizens. This chapter summarizes an attempt to meet this need by applying the methodology of Resilience Engineering (Hollnagel, Woods and Leveson, 2006; Hollnagel et al., 2011) and the concept of Safety-II (Hollnagel, 2012, 2013). First, official accident reports are reviewed to identify the key factors that contributed to the accident and its negative after-effects. Then, an additional in-depth review is carried out to obtain a structured description of the causal relationships and to identify so-called second stories describing more fundamental factors that eventually contributed to the Fukushima disaster.

### *Target of Study*

Subsequent to the disaster, several accident analysis committees were organized in Japan. The Investigation Committee on the Accident at Fukushima Nuclear Power Stations of TEPCO (also called the Hatamura Committee), and The Fukushima Nuclear Accident Independent Investigation Commission organized by The National Diet of Japan (also called the Kurokawa Commission) were the most influential committees since they were established by the Japanese government and the National Diet of Japan, respectively. Another committee named the Independent Investigation Commission on the Fukushima Daiichi Nuclear Accident (also called the Kitazawa Commission) was established by an organization named Rebuild Japan Initiative Foundation (RJIF). After highly intensive investigations, all these groups have published reports, which are herein called the Hatamura report (Hatamura, 2012), the Kurokawa report (Kurokawa, 2012) and the Kitazawa report (Kitazawa, 2012), respectively. The present study is based on in-depth analysis and examination of these three reports. The report issued by The American Nuclear Society Special Committee on Fukushima (Klein and Corradini, 2012) is also reviewed in this study for purposes of comparison.

### *Focus of Study*

These reports cover a large number of factors, which contributed to the extremely severe accident. Though the contents differ to some extent, they all contain strong criticism of TEPCO, the regulatory bodies such as the Nuclear and Industrial Safety Agency (NISA) and the Nuclear Safety Commission (NSC), the Cabinet of Japan, and the Japanese nuclear community. The focus of this chapter is placed on major causes of the accident commonly identified and on the recommendations derived from the identified causes. In addition to the key findings and recommendations described in the reports, narrative statements referred to in the reports as testimonies by witnesses and interviewees are carefully reviewed.

This approach is adopted because the narrative statements are sometimes more informative since they are less modified by subjective views of the interviewers.

## Overview of Investigation Reports

### *Findings and Recommendations*

The investigation by various bodies was undertaken by focusing on the activities of involved individuals and organizations during the accident, but then the focus was naturally extended to activities prior to the accident. The time span of the retrospective investigation was extended to include historical efforts concerning seismic Probabilistic Safety Assessment (PSA) and severe accident management in the early 1990s. Some remarks on lessons learned from the accident at the Three Mile Island (TMI) nuclear plant, which occurred in 1979, were also made. As typical examples of major findings, the ones given in the Hatamura report (Hatamura, 2012) are compiled from the author's viewpoint and presented in Table 4.1.

Issues 1 through 4 essentially imply that a number of improvements must be made in order to enhance defense-in-depth capabilities of the NPSs. The defense-in-depth capability implies that the NPS is not only able to prevent occurrence of anomalies and accidents, but also able to mitigate consequences of an accident if it happens in spite of the preventive measures. The capability also implies that proper plans must be prepared in advance for effective evacuation of residents if it becomes necessary in spite of the mitigation efforts. Issues 5, 6 and 7 are related to crisis management, including crisis communication. These issues are also important constituents of defense-in-depth capability. Issue 8 indicates the need for rebuilding safety culture, and issue 9 implies the need for further investigation. Recommendations in the Hatamura report to address the issues presented in Table 4.1 are given in Table 4.2.

## Table 4.1 Major issues mentioned in the Hatamura Committee report

| Issue No. | Headings | Key statements |
|---|---|---|
| 1 | Building of fundamental and effective disaster preventive measures | Quite a number of problems exist .... These problems should be reviewed and resolved ... In doing so, concerned organizations should sincerely take into consideration the recommendations the Investigation Committee has made and they should do so with accountability to society for its process and results. |
| 2 | Lack of viewpoint of complex disasters | Risks of a large-scale complex disaster should be sufficiently considered in emergency preparedness. |
| 3 | Change needed in an attitude to face risks | It is necessary to humbly face the reality of natural threats, diastrophism and other natural disasters ... |
| 4 | Importance of "deficiency analysis from the disaster victims' standpoint" | If nuclear operators and regulatory bodies overestimate the safety of "system core domain" ... safety measures will fail. Safety measures in the "system support domain" and "regional safety domain" need to be able to function independently in the case of an emergency. |
| 5 | The issue of "beyond assumptions" and lack of the sense of crisis at the administrative bodies and TEPCO | Scientific knowledge of earthquakes is not yet sufficient yet. The latest research results should be continually incorporated in emergency preparedness. |
| 6 | Issues of the government crisis management system | The crisis management system for a nuclear emergency should be urgently reformed. |
| 7 | Issues of the provision of information and risk communication | It is necessary to build mutual trust between the public and the government and to provide relevant information in an emergency while avoiding societal confusion and mistrust. |
| 8 | Importance of a safety culture vital to the lives of the public | In view of the reality that safety culture was not necessarily established in our country, the Investigation Committee would strongly require rebuilding of safety culture ... |
| | Necessity of continual investigation of the whole picture of accident causes and damage | Items that were not subjected to investigation and verification by the Investigation Committee remain of great importance to the victims. |

**Table 4.2 Recommendations mentioned in Hatamura report**

| Rec No. | Headings | Target issue |
|---|---|---|
| 1 | Recommendations for a basic stance for safety measures and emergency preparedness | Emergency preparedness in light of complex disasters; Changing an attitude to face risks; Incorporating the latest knowledge in the emergency preparedness. |
| 2 | Recommendations for safety measures regarding nuclear power generation | Necessity of comprehensive risk analysis; Severe accident management. |
| 3 | Recommendations for nuclear emergency response systems | Reforming the crisis management system for a nuclear emergency; Nuclear emergency response headquarters. |
| 4 | Recommendation for damage prevention and mitigation | Provision of information and risk communication; Improvement of radiation monitoring operations and SPEEDI system; Evacuation procedures of residents; Intake of stable iodine tablets; Improvement of medical care institutions; Public understanding of radiation effects. |
| 5 | Recommendations for harmonization with international practices | International practices such as IAEA safety standards. |
| 6 | Recommendation for relevant organizations | Reforming nuclear safety regulatory body and TEPCO; Rebuilding a safety culture. |
| 7 | Recommendations for continued investigation of accident causes and damages | Continued investigation of accident causes; Extended investigation of the whole picture of accident damage. |

Recommendations 1 and 2 are relevant to the activities prior to the accident with emphasis on emergency preparedness, and recommendations 3 and 4 are relevant to the activities during the accident such as crisis management and communication. Recommendations 5 and 6 are more fundamental issues related to basic safety culture and international harmonization. Recommendation 7 reflects the recognition of the Investigation Committee that the investigation is still in its interim stage.

**Needs for In-depth Analysis**

The findings and recommendations given in the report contain quite broad viewpoints adopted by the Investigation Committee. These recommendations seem to be valid in general. The same is true for other reports such as the Kurokawa report and Kitazawa report. However, several drawbacks concerning the findings and recommendations can be identified as detailed below:

- The findings and recommendations must be structurally organized and simplified: The findings and recommendations have been derived via examination of a wide variety of difficulties experienced during the accident in diverse spatial domains and time periods. In other words, the recommendations seem to reflect a linear model of causality corresponding to each difficulty experienced along with the progression of the accident. As a natural outcome of this approach, the contents of Table 4.1 and Table 4.2 are quite complicated. The itemized list of recommendations in Table 4.2 is actually even more complicated because each of the recommendations implies multiple requirements. According to the widely acknowledged concept of Occam's razor (Rissanen, 1978), also known as law of parsimony (Akaike, 1974), out of multiple competing theories or explanations, the simplest one is to be preferred. This principle is also expressed as "Entities are not to be multiplied beyond necessity." Further efforts must be made to find more organized and integrated representations of the causal relationships. In this regard, an attempt must be made to compile and express the complicated outcomes of the investigation in a simpler fashion.
- Second stories must be clarified: The messages in the reports seem to be reasonable as far as "first stories" (Woods and Cook, 2002) are concerned. For example, in Table 4.1, under the heading of "Lack of viewpoint of complex disasters", it is

stated that "Risks of a large scale complex disaster should be sufficiently considered in emergency preparedness." Also, in Table 4.1, under the heading of "Issues of the provision of information and risk communication", it is stated that "It is necessary to build mutual trust between the public and the government and to provide relevant information in an emergency while avoiding societal confusion and mistrust." It is obvious that these statements are reasonable. However, a question, that is, why the responsible organizations such as TEPCO and NISA failed to meet such necessary conditions prior to the accident, is not explicitly dealt with. More efforts must be made to pursue second stories (Woods and Cook, 2002) beneath the surface to discover other contributors and/or causes. Otherwise, the recommendations can have only limited effectiveness.

- Accountability for "safety" must be pursued: Whatever countermeasures are implemented, it is obviously necessary that the implication of attained "safety" be accountable, as emphasized in issue 1 of Table 4.1. In other words, the consequence of implementation of countermeasures must be transparently explained and the implication of resultant "safety" must be clarified. Otherwise, the resultant improvement in safety will be unacceptable to the public.

An attempt to meet the above-mentioned requirement s has been conducted in light of Resilience Engineering. Observations derived from the attempt are described in the next section.

## Restructuring the Findings and Recommendations

The activities of people involved in the Fukushima accident have been reviewed in terms of the four essential capabilities that together define resilience: responding, monitoring, anticipating and learning (Hollnagel, Woods and Leveson, 2006; Hollnagel, Woods, Paries and Wreathal, 2011). This approach has been adopted with an expectation that the diverse activities conducted during the accident can be structured in a simpler manner.

### *Analysis of Situation Prior to the Accident*

As is widely known, the ability to learn was unacceptably poor in TEPCO, NISA, and NSC. The issue of severe accident management had been discussed and studied in Western countries and a number of documents had been published. It is highly likely that Japanese utilities had various opportunities to examine and learn from the documents. In addition to learning from such documents, the Japanese nuclear community could have learned from actual accidents/incidents. On December 27, 1999, an unexpectedly strong storm flooded the Blayais Nuclear Power Station in France, resulting in water damage of pumps and containment safety systems. Also on December 26, 2004, the Sumatra tsunami attacked the Madras Nuclear Power Plant in India, resulting to an emergency shutdown due to tsunami-induced damage to a seawater pump. The Japanese nuclear energy organizations could have obtained informative lessons from these events.

In conjunction with the poor ability to learn, the ability to anticipate was also poor. Since the nuclear energy organizations were preoccupied by a mindset that an extremely large earthquake and tsunami were highly unlikely to happen, they made practically no effort to anticipate and prepare for these external events and resultant severe accidents. This mindset is clearly evidenced by repeated neglect of warnings raised by various actors.

It is now clear that the NISA had received several warnings from researchers in relevant academic areas concerning the possibility of a gigantic tsunami in Fukushima and adjacent prefectures. Another warning concerning a possible tsunami had been raised by Diet member H. Yoshii in a budget committee meeting of the Diet. He had also addressed the possible loss of external AC electricity and station blackout. These warnings made by H. Yoshii were issued in 2006. Similarly to the TMI accident, the Fukushima accident could have been avoided, or considerably mitigated, if these warnings had been properly taken into consideration. Needless to say, the lack of ability to learn and anticipate is in contrast to the requirement of "a constant sense of unease" mentioned as an important necessary condition to establish a resilient system (Hollnagel, 2006b)

*Analysis of Situations During the Accident*

Since the abilities to learn and anticipate were poor, the ability to respond to the tsunami and subsequent severe accident was terribly deficient. Most of the resources needed for utilization during the accident were unavailable. TEPCO people collected 12V batteries from cars and buses and used them to activate the instrumentation systems for measuring important plant parameters such as reactor pressure and water level. But the number of available batteries was far fewer than needed. In addition, the majority of response actions were carried out in the "scrambled" (Hollnagel, 1993) mode. One typical example is the ignored response to the order issued by the Director of Fukushima-Daiichi NPS to prepare for using fire engines to inject cooling water in the reactor. Though the verbal order was recognized by personnel in the emergency response center, nobody actually responded to the order since no division had been assigned in advance to conduct such an unusual way of accident management.

The ability to monitor was also deficient. As with the weakness of the ability to respond, the monitoring activities were conducted in the "scrambled" mode. One of the worst failures in monitoring resulted from the invalid belief that the isolation condenser (IC) was functioning in unit 1 reactor. Since nobody foresaw the possibility of IC malfunctioning, which was caused by an unintended closure of valves in the piping connecting the pressure vessel to the IC, the unit 1 reactor was left uncooled, resulting to a hydrogen explosion. If somebody in TEPCO had considered the possibility of IC malfunctioning and tried to monitor the status of the IC and other reactor cooling systems, the situation could have been significantly improved.

Another example of serious failure in monitoring was the weak attention to the condition of the unit 2 reactor. More attention had been paid to prevention of a hydrogen explosion in the unit 3 reactor because of the incorrect assumption that the situation of unit 2 was less dangerous. In reality, that was not true. When the pressure and water temperature inside the suppression chamber were measured by tentatively using batteries, the measured values indicated that the condition of unit 2 was becoming dangerous.

If monitoring of the plant condition had been carried out several hours earlier, the situation could have been much improved. It should be noted that the majority of radioactive material released from the Fukushima-Daiichi NPS was not caused by the hydrogen explosions of unit 1 and unit 3, but rather by the break of pressure boundary of unit 2.

Based on the above-mentioned analysis, it can be concluded that the accident is attributable to deficiencies in each of the four essential capabilities. The large number of findings and recommendations in the investigation reports can be restructured in a more organized manner corresponding to the four capabilities. Also, it is obvious that the most glaring deficiency in the four capabilities is the poor learning, which inevitably led to the poor preparedness for severe accidents. It is possible to argue further and criticize the organizational and managerial flaws in TEPCO, NISA, and other related organizations. The criticisms are valid per se, and thus the flaws must be remedied. However, it should be stressed that the accident could have been prevented or at least significantly mitigated if the learning-based preparedness had been in a better condition.

### *Second Stories*

The restructured version of findings and recommendations given in the preceding section provides us with a clearer view of the accident causes and contributors. However, the improved view is still dependent on first stories. Deeper interpretations based on other fact-finding efforts are certainly needed.

First, the question, that is, why responsible organizations such as TEPCO and NISA failed to meet such necessary conditions prior to the accident, must be resolved. In Table 4.1, issue four states as follows: "If nuclear operators and regulatory bodies overestimate the safety of 'system core domain' … safety measures will fail."

The Hatamura report also stated that the organizations' ignorance and/or neglect of the possibility of a tsunami had been caused by lack of imagination and complacency within the organizations. But this interpretation seems to be superficial. A deeper interpretation can be derived from a description provided in another investigation report (Kitazawa, 2012):

> Multiple members of TEPCO management board stated, "I doubted that safety measures of TEPCO NPSs were sufficient. But I was reluctant to express my concern since I felt my opinion was minor in the board."

This description clearly indicates that a minority of the board members had concerns about a tsunami and severe accident management but stayed silent because of a lack of confidence in communicating their concerns about the safety of the NPS.

A similar issue is related to the fear of lawsuits. The Kurokawa report criticized TEPCO and NISA for having been reluctant to improve their preparedness against severe accidents because of a fear of lawsuits. This interpretation is again superficial. The risk of lawsuits has in fact been high. To prepare for lawsuits is of course quite demanding for utilities and NISA. Most utilities operating NPSs have experienced anti-nuclear lawsuits, and several lawsuits are still ongoing. It is likely that the accusers would strongly claim the safety measures currently implemented at the NPSs have proved to be insufficient if the accused utility decided to implement additional measures in preparation for severe accidents. Though such a claim is not valid from the viewpoint of defense-in-depth principle, it is often convincing to jurists and the concerned public. This dilemma also originates from the difficulty in communicating the safety of NPSs.

Another issue is the risk of long-term shutdown magnified by the shared regulatory authority. The comments of Nobuaki Terasaka, the director of NISA at the time of the Fukushima accident, and Atsuyuki Suzuki, who served as the Chairman of NSC from 2006 to 2010, both quoted in the Hatamura report, stressed the difficulty of explaining the need for measures for severe accident management to local government and the public. This issue is also mentioned in the ANS Committee Report (Klein and Corradini, 2012) as follows:

> One of the key lessons learned after the TMI-2 accident was to reform and strengthen the independence and technical competence of the NRC (Nuclear Regulatory Commission). Many other nations followed. However, Japan did not change its regulatory governance because to do so would centralize too much authority in its central government, which would upset the shared authority arrangement with the prefectural government.

Because of the shared authority arrangement, TEPCO, and other utilities as well as NISA, always need to make considerable effort to obtain the agreement of the prefectural government to restart operations after every scheduled and unscheduled shutdown. Their reluctance to implement additional measures for severe accident management should be interpreted in this context also. In short, all the second stories described in this section are attributable to the difficulty in communicating the safety of NPSs in conjunction with the defense-in-depth principle.

It should be noted that a considerable portion of nuclear risk is socially constructed (Kitamura, 2009). As described above, the difficulty in communicating safety based on the defense-in-depth principle to top-management people in a company, to juries and the public, and to local authorities eventually led to the lack of preparedness for tsunamis and severe accidents. The messages from the investigation reports would have been more convincing and prescriptive (that is, resolution-oriented) if this aspect had been articulated more explicitly.

### *Accountability, Safety-I, Safety-II, and Resilience Engineering*

It is definitely necessary to resolve the difficulty experienced in communication of nuclear safety. Through experiences of public communication concerning nuclear technology for more than ten years (Yagi, Takahashi, Kitamura, 2006), the author believes that the fundamental difficulty lies in the confused recognition of nuclear safety, and safety in general. The people in nuclear organizations have some understanding about the concept of defense-in-depth, but they are not sufficiently knowledgeable to provide a convincing response to the accusation from anti-nuclear activists that any attempt to implement add-on measures for severe accident management is clear evidence that the current state of the NPS is not absolutely safe.

The author believes that the first necessary step to resolve the difficulty is to provide and share the idea of Safety-II (Hollnagel, 2012, 2013). The traditional definition of safety is termed Safety-I, where the purpose of managing safety is to attain and maintain the state where the number of adverse events is controlled to be as low as reasonably achievable. Safety-I practices presume that the target system and operation environment can be completely understood and specified. Safety-I also assumes that safety can be achieved by exhaustively eliminating the causes of adverse events. As a result, a state of safety can be achieved by compliance to rules and procedures. In this context, the explanation of the need for additional measures for severe accident management is logically difficult. Similarly, denying the accusations of anti-nuclear activists is also difficult.

On the contrary, an alternate concept of safety, that is, Safety-II, is defined as a state where "things go right." Safety-II also assumes that the target system and its environment are subject to uncertainty and unforeseen disturbances. Within the framework of Safety-II, the expected performance of the target system is not just to maintain a static condition, but also to override a large disturbance. Furthermore, when the disturbance is so large that the system performance is significantly degraded, the system is expected to recover from the degraded state as smoothly as possible. Safety-II logically includes Safety-I since something that goes right cannot go wrong at the same time.

It should now be clear that Safety-II is consistent with the concept of defense-in-depth, where what is pursued is not Safety-I but Safety-II. It should also be clear that the four main capabilities, the preparedness of resources and the constant sense of unease introduced in the framework of Resilience Engineering are excellent as practical guidelines to materialize the concept of defense-in-depth. The introduction of the concept of Safety-II, renewed understanding of the concept of defense-in-depth, and the methodology of Resilience Engineering for materialization of Safety-II together would enable nuclear experts to acquire communication capability to explain the reality of nuclear safety with accountability to society and concerned citizens.

## Conclusion

The reports issued by the official investigation committees have been reviewed in this chapter. It should be clear that in their reports, these committees have issued recommendations for eliminating the causes of the adverse events experienced during the accident. The basic methodology behind the investigation is to find out a linear model of causality that corresponds to each of the adverse events, and to propose countermeasures to nullify the causal relationship. This methodology is also regarded as a scenario-driven, descriptive approach since the adverse events and the related models of causality are identified along with the scenario of the actual accident. The present review has been conducted from a different viewpoint, namely, that lessons for enhancement of nuclear safety can be identified by looking at the accident from a "what-if" perspective. This is equivalent to conducting a thought experiment where the crucial functions proposed by Resilience Engineering are applied to critical time points prior to the accident. Although this approach might be regarded as just a hypothetical exercise, the results are sufficiently informative to provide us with useful guidelines for significant enhancement of nuclear safety in the future. Among the four main functions, proactive and continuous application of learning and anticipating, in particular, will be beneficial to avoid the recurrence of severe nuclear accidents. The observations obtained through the present attempt clearly demonstrate the high potentiality of the Resilience Engineering in providing a systematic way toward safety improvements with accountability to society. Additional potentiality of Resilience Engineering as a methodology for resolving conflicts between technology and society should be acknowledged because public participation in technical conflict resolution is becoming a standard procedure in modern society, because a considerable portion of the risk of a large-scale socio-technical system is regarded as socially constructed (Kitamura, 2009), and because technological activities must be conducted with accountability to the public.

## Commentary

The more serious an accident is, the more urgent is the need to explain what has happened, usually by pointing to some clearly recognizable causes. While the several investigation committees that have looked at the Fukushima Daiichi disaster were no exception to the rule, this chapter tries to go beyond the multiple first stories by applying the principles of Resilience Engineering, and by looking at what happened from a safety perspective. Large-scale socio-technical systems cannot be made safer by addressing specific causes one by one and by filling identified gaps with even more technology. Large-scale socio-technical systems are socially constructed and their ability to succeed cannot be strengthened unless the methods that are used can account for that.

# Chapter 5
# Criteria for Assessing Safety Performance Measurement Systems: Insights from Resilience Engineering

Tarcisio Abreu Saurin, Carlos Torres Formoso
and Camila Campos Famá

Although the use of safety performance measurement systems (SPMS) is an important practice, the literature does not offer clear recommendations on how they might be assessed. In this chapter, six criteria for assessing SPMS are proposed, which are founded on the resilience engineering paradigm. The use of those criteria is illustrated by two case studies, in which the SPMS of two construction companies were evaluated. The insights obtained from using those criteria are unlikely to result from general criteria for assessing performance measurement systems.

## Introduction

Performance measurement is closely related to the four abilities of resilient systems proposed by Hollnagel (2009): responding, monitoring, anticipating and learning. Indeed, it can support the identification of disturbances the system should respond to as well as producing data that enable learning to take place systematically. The connections between performance measurement and the monitoring and anticipating abilities are straightforward, since the former may be interpreted as equivalent to data collection and the latter should be a consequence of analysing the data produced.

A number of studies have discussed performance measurement, considering ideas from the RE perspective (for example, Wreathall, 2011), including the proposal of methods for developing resilience-based early warning indicators (Oien et al., 2011). The aim of this chapter is to propose a set of criteria for assessing safety performance measurement systems (SPMS) from the RE perspective, which are intended to complement general criteria that any performance measurement system should meet (Neely et al., 1997). This set of criteria was applied in two case studies, carried out in the construction industry. Although a number of studies on RE place an emphasis on complex socio-technical systems that involve intensive use of automation, such as aviation and power plants, previous studies have shown the benefits of applying RE to the construction industry (Saurin et al., 2008). Indeed, construction is recognized for having typical characteristics of complex systems, such as uncertainty and a large number of dynamically interactive elements. RE principles and methods are particularly suitable for this type of environment, in which human performance variability is frequent and necessary for successful performance (Hollnagel, 2012).

## Criteria for Assessing SPMS from the RE Perspective

As the RE abilities are generic for any aspect of safety management, they need to be translated into other constructs, which can support the operationalization of RE for the purpose of assessing SPMS. The proposed set of criteria for assessing SPMS is focused on the contents and efficiency of the SPMS. Of course, organizational learning, as well as the implementation of actions arising from the insights provided by the criteria, are required if they are to make a real contribution to safety performance. The proposed criteria are:

- **(a) The SPMS should monitor normal work**: this criterion arises from a distinctive premise of RE, namely the acknowledgment that accidents are not intrinsically different from normal performance. 'Normal' should be interpreted in the sense of everyday work (Hollnagel, 2012), with all shortcuts that gradually become incorporated in the routine and are accepted as normal, at least until a serious accident happens. Monitoring of normal work should shed light on the reasons for performance variability, since this supports its management (Macchi, 2010).

- **(b) The SPMS must be resilient**: this criterion arises from the dynamics of a complex system, which implies that the SPMS should have the ability to adapt in order to continue capturing relevant information (Oien et al., 2011). In order to check if and how the SPMS is adapting, it should also be monitored, such as, for instance, by means of external audits and metrics to assess its efficiency and effectiveness. Indeed, a non-resilient SPMS may deteriorate (for example, when people stop collecting and analysing data) and become obsolete.
- **(c) The SPMS should monitor hazards throughout the socio-technical system**: a source of hazard might be fairly distant, in time and space, from the situations it affects. RE recognizes this fact, once it proposes that the combination and propagation of the normal variability of multiple functions, over time and space, can lead to unexpected outcomes (Hollnagel, 2009). Thus, a broad scope of hazard identification and monitoring should be adopted, comprising the technical, social, work organization, and external environment dimensions of a socio-technical system.
- **(d) The SPMS should get close to real-time monitoring**: due to the dynamics of a complex system, the information provided by SPMS might become out-of-step in relation to the system's status (Hollnagel, 2009). Thus, when performance information is delivered to its users, the system might no longer be as it was at the time of data collection. This criterion implies that there should be reduced the time lag between the events and data analysis. A possible alternative to shorten this lag is the decentralization of the tasks of collecting, analysing and disseminating information generated by the SPMS. In fact, distributed control is typical of complex systems, and the design of SPMS should take advantage of this fact.

- **(e) The SPMS should take into account performance in other dimensions of the organization**: this criterion is derived both from the RE premise that safety cannot be separated from business and from criterion (c). As a result, the SPMS should permeate all areas and activities, not only those normally associated with safety (Hopkins, 2009). Based on this criterion, it can be assumed that the other performance measurement systems, such as quality and environment control, may indirectly provide important information for SPMS.
- **(f) The SPMS should balance the trade-off between completeness and ease of use**: due to constraints of resources and time, human performance cannot maximize efficiency and thoroughness at the same time (Hollnagel, 2012). A SPMS is not immune to the efficiency-thoroughness trade-off. From the perspective of completeness (or thoroughness), the evaluation of safety in complex systems cannot focus on few indicators and subjects, otherwise it will not capture the nuances that comprise the situation. From the perspective of ease of use, which is an aspect of efficiency, any performance measurement system must be cost-effective. The use of an underlying safety paradigm, such as RE, helps to manage this trade-off, to the extent that there will be guidelines on what is important to measure and what is not.

Four out of the six criteria presented above are innovative, in comparison with either general criteria for assessing performance measurement systems or safety management approaches limited to compliance with regulations. They are: SPMS should monitor normal work; SPMS should monitor hazards throughout the socio-technical system; SPMS should get close to real-time monitoring; and SPMS should take into account performance in other business dimensions. Moreover, these four criteria are fairly safety-specific, as it is not obvious how they could be relevant for other business dimensions.

The other two criteria might be portrayed as reinterpretations of existing criteria from the RE perspective. The criterion stating that the SPMS should be resilient resembles the well-known recommendation, for performance measurement systems in general, that they should be themselves an object of continuous improvement (Neely et al., 1997). However, the RE perspective provides guidance on which grounds improvement should take place, such as by learning from normal work and taking into account performance on business dimensions that are not directly associated with safety.

The criterion stressing the need for the SPMS balancing the trade-off between completeness and ease of use resembles the advice that any performance measurement system should be cost-effective (Neely et al., 1997). However, the proposed criterion translates the cost-effectiveness trade-off in a more meaningful language for an SPMS. Indeed, while ease of use has an impact on cost, completeness has an impact on the SPMS effectiveness. Moreover, the RE perspective supports the identification of what is worth measuring by an SPMS, which is an issue neglected by generic recommendations without an underlying safety paradigm.

Also, while general criteria state that the metrics should be aligned with strategies (Neely et al., 1997), it is not straightforward to analyse this alignment when the focus is on safety. In this study, it is argued that both a company strategy and a safety management paradigm share an important commonality, as they provide a vision of what a company should be. In this respect, the proposed criteria provide a basis to assess the extent to which an SPMS is aligned with the RE view, which may be useful for companies that pursue the implementation of RE principles.

## Research Method

Two construction companies (A and B) in Brazil were selected for the case studies. Company A's main activities were the development and construction of buildings projects for middle- and higher-middle-class customers or residents, and it had about 1,200 employees. Company B was mostly focused on complex and fast hospital and industrial building projects. It had about 200 employees. The safety management systems of both companies were quite similar, as they had a number of noteworthy practices in common, such as a full-time safety specialist in all building sites and weekly construction planning meetings, involving both production and safety staff. Both companies had safety indicators that were not limited to those required by regulations, and had been implemented in a fairly standardized way in most of their building sites.

As the criteria for assessing an SPMS from the RE perspective were fairly abstract, it was necessary to break them down into sub-criteria which could guide data collection (Table 5.1). Two sites of each company, which were considered to be typical for their SPMS, were chosen for collecting data. In each site, a member of the research team carried out four visits over a period of two months, each of them lasting approximately two hours. In the first visit, the aim was to obtain an overall understanding of the SPMS, emphasizing the identification of both the metrics adopted and of the procedures of data collection, analysis and dissemination. In the three remaining visits, doubts arising from the first visit were clarified and the investigation moved towards more specific topics, as the researchers looked for the sources of evidence listed in Table 5.1. After concluding data collection, the researchers prepared a report on the evaluation of the SPMS, which, in both companies, was discussed with production and safety staff.

**Table 5.1 Criteria and sub-criteria for assessing SPMS from the RE perspective**

| Criteria | Sub-criteria | Sources of evidence |
|---|---|---|
| **1.** The SPMS should monitor normal work | 1.1 The SPMS generates information that is based on analysing normal work (i.e. everyday performance), instead of only that which arises from analysing failures and adverse events<br>1.2 The SPMS generates information on the sources and reasons for performance variability that leads to successful outcomes and variability that leads to unsuccessful ones | Handbooks and forms used for describing the indicators and collecting them<br>Reports with the results of the indicators, accident investigation reports<br>Semi-structured interviews with operational teams, staff responsible for the SPMS and intermediate level staff<br>Observation of construction tasks, emphasizing the search for practices not covered by procedures and regulations |

**Table 5.1** *Continued*

| | | |
|---|---|---|
| **2.** The SPMS should be resilient | 2.1 The procedures for collecting, analysing and disseminating the metrics evolve over time<br>2.2 The metrics are excluded, adapted or included, as a result of changes in the risks or enhancement of the SPMS<br>2.3 There are mechanisms for evaluating the effectiveness and efficiency of the SPMS | Handbooks and forms used for describing the indicators and collecting them<br>Semi-structured interviews with staff responsible for the SPMS<br>Observations of formal and informal events in which the results of the indicators are discussed |
| **3.** The SPMS should monitor hazards throughout the socio-technical system | 3.1 A broad scope for identifying hazards is adopted (e.g., organizational pressures, process safety hazards, personal safety hazards, health hazards, side effects of incentive programs linked with safety performance, etc.)<br>3.2 The SPMS is concerned with the whole life-cycle of the socio-technical system, from its design to its dismantling/replacement | Handbooks and forms used for describing the indicators and collecting them<br>Observations of formal and informal events in which the results of the indicators are discussed<br>Semi-structured interviews with staff responsible for the SPMS |
| **4.** The SPMS should get close to real-time monitoring | 4.1 Feedback is provided, to the interested parties, quickly after safety relevant information is collected and analysed<br>4.2 The tasks of collecting, analysing and disseminating information are distributed over several agents (e.g., front-line workers, supervisors, managers), reducing dependency on centralized and possibly overloaded control mechanisms | Handbooks and forms used for describing the indicators and collecting them<br>Observations of formal and informal events in which the results of the indicators are discussed<br>Semi-structured interviews with staff responsible for the SPMS |
| **5.** The SPMS should take into account performance in other business dimensions | 5.1 The SPMS assesses how safety is performing in comparison with other business dimensions, providing insights on the extent to which safety has been valued<br>5.2 The indicators not directly connected with safety (e.g., those of cost, time and quality) are interpreted from a safety perspective | Handbooks and forms used for describing the indicators and collecting them<br>Handbooks and forms related to indicators not directly connected to safety<br>Observation of meetings in which the indicators not related to safety are discussed<br>Semi-structured interviews with staff responsible for the SPMS and for other areas of performance measurement |

**Table 5.1 *Concluded***

| | | |
|---|---|---|
| **6.** The SPMS should balance the trade-off between completeness and ease of use | 6.1 The resources (human, technical and financial) needed to maintain the SPMS operational are adequate<br>6.2 The design and routines for operating the SPMS are fully understood by those responsible for its management<br>6.3 The SPMS takes advantage of the qualitative data that is collected to calculate the indicators, in order to provide a richer and accurate view of safety performance<br>* Other insights concerning the SPMS completeness can be drawn from criterion 3 (The SPMS should monitor hazards throughout the socio-technical system) | Semi-structured interviews with the staff responsible for the SPMS<br>Semi-structured interviews with staff not directly involved with the management of the SPMS (e.g., workers and supervisors)<br>Observations of the procedures for collecting the indicators and disseminating their results<br>Observations of formal and informal events in which the results of the indicators are discussed |

## Results

*Main Characteristics of the SPMS of the Companies Investigated*

Table 5.2 presents the main characteristics of each company SPMS. These characteristics indicate a concentration of tasks on the safety specialists, and a very low involvement by top managers and production managers, especially in company A.

**Table 5.2 Main characteristics of the SPMS of companies A and B**

| Characteristics of the SPMS | Company A | Company B |
|---|---|---|
| How many indicators? Were these collected in all sites? | Eight indicators, which were collected in all sites | Seven indicators, which were collected in all sites. Five of them were similar to those collected in company A |

**Table 5.2** *Concluded*

| | | |
|---|---|---|
| Who did collect and process data for calculating indicators? | The safety specialists, although foremen, production managers and workers provided some of the information required by the specialists | The same as company A |
| How often the indicators' results were produced and a formal report was generated? | There was a monthly compilation of data from each site and a monthly compilation using data from all sites | The same as company A |
| How did report look? | Both the individual report from each site and the general report had graphs and comments on them. There were also visual alerts, by using colours to indicate the status of the indicator in relation to the targets | There were reports, graphs and comments on them, but no visual alerts as in company A |
| Who was involved in the discussion of the indicators' results? | Only members of the safety staff, in a monthly meeting | Safety staff, production staff and top managers, in a monthly meeting |
| Main moments of analysis and onward transmission of each indicator's results | The above-mentioned monthly meeting of the safety staff; the monthly meeting of a committee that is mandatory by regulations; and the daily safety training meetings, coordinated by the safety specialist. In each site, there was also a board that showed the monthly results of the indicators | The above-mentioned meeting for discussing the results, the monthly meeting of a safety committee that is mandatory by regulations; and the daily safety training meetings, coordinated by the safety specialist |

*Safety Indicators Used by the Companies*

Five indicators were used in both companies. While three of them are fairly straightforward (accident frequency rate, near misses frequency rate and index of training, which measured the amount of training provided to workers), clarification is necessary for two other indicators. The Percentage of Safety Activities Concluded (PSAc) indicator was inspired by another indicator adopted in both companies, called Percentage of Production Activities Completed

(PPAC). The PSAc formula measures the ratio between the number of safety activities concluded and the number of safety activities planned. Although there was no formal definition of what was regarded as a safety activity, the observations indicated it was concerned with implementing physical protections (for example, guardrails) and access equipment to work stations, such as ladders. The PSAc was monitored weekly, and the causes that led to the non-completion of safety activities were discussed. The aim of the NR-18 index (INR-18) was to evaluate the compliance with the main Brazilian regulation concerned with construction safety, called NR-18. The INR-18 was calculated from a checklist that had 213 items, corresponding to the ratio between the total of items marked with yes (meeting the regulation) and the total of items marked with yes or no.

Three indicators were used only in company A: estimate of fines due to non-compliance with NR-18; the number of production stoppages due to lack of safety; and the index of subcontractors' performance, which took into account a number of requirements applicable to the subcontractors. Two indicators were used only in company B: frequency rate of first-aid accidents; and index of compliance and commitment, which assessed the number of safety notifications solved within the deadline established by the safety specialist, who carried out a daily inspection of site activities.

## Evaluation of the SPMS Based on the Proposed Criteria

### *The SPMS Should Monitor Normal Work*

Five out of the ten indicators existing in both companies (accident frequency, near miss frequency, estimate of fines, frequency of first aid accidents, number of stoppages due to lack of safety) are focused on measuring adverse events. Rather than monitoring normal work (sub-criterion 1.1), they monitor more or less unfrequent events that reflect lack of safety, instead of its presence. Nevertheless, those indicators could give insights into normal work, provided the descriptions of adverse events were compared with the work prescribed. This analysis could reveal adaptations that have been incorporated into the day-to-day routine.

The other five indicators (index of training, index of compliance with NR-18, index of subcontractors performance, percentage of safety activities completed and index of compliance and commitment) are focused on analysing normal work and they monitor either the presence of safety or actions that have been adopted to create safety, such as training and planning. However, the index of compliance and commitment is the only indicator that involves observing people working, which is necessary for identifying human performance variability, a key concern from the RE perspective. Indeed, the other indicators might be limited to the observation of the technical system. Although information on the sources and reasons of variability (sub-criterion 1.2) could be extracted from the ten indicators existing in both companies, such extraction was limited to the sources and reasons of variability leading to unsuccessful outcomes. Indeed, when analysing data arising from all indicators, the staff of both companies focused on identifying what went wrong and why, rather than identifying what went right and why. Data collected by the researchers showed that learning opportunities were missed due to this approach, since successful outcomes were not necessarily due to adherence to formal system design.

*The SPMS Should Be Resilient*

The evaluation according to this criterion was hindered by the duration of the case study. In fact, two months were not enough to detect substantial changes in the procedures for collecting, analysing and disseminating the metrics (sub-criterion 2.1) as well as to detect major changes in the metrics, such as inclusions, exclusions, or adaptations (sub-criterion 2.2).

The resilience of the SPMS could also be enhanced based on insights arising from formal assessments of their effectiveness and efficiency (sub-criterion 2.3). However, none of the companies had procedures for evaluating their SPMS, relying mostly on gut feelings of the safety staff. Nevertheless, the SPMS provided a large amount of information, which, if properly interpreted, could be used for such assessments, in order to reduce the dependence on safety staffs' individual insights. For example, the pieces of information used to calculate five indicators (frequency rate of near misses, number of stoppages, accident frequency rate, index of compliance and commitment and first-aid) could potentially point out hazards of any nature. Therefore, those indicators could be interpreted as meta-monitoring mechanisms, since they provided data for monitoring the SPMS itself. In companies A and B, for example, the need for the indicator called index of training could be questioned, as the lack of training was not identified as a major contributing factor to accidents, near misses and non-execution of safety activities.

*The SPMS Should Monitor Hazards Throughout the Socio-technical System*

Even though eight indicators existed in company A, and seven in company B, the implicit definition of what counted as a hazard (sub-criterion 3.1), and should therefore be monitored by the SPMS, was limited. Some indicators had a narrow focus on certain elements and hazards of the socio-technical system, such as the index of compliance with NR-18, which detected failures related to the technical system, that is, whether physical protections were installed and kept in good conditions. By contrast, other indicators, as mentioned in the previous section, could potentially monitor a broader range of hazards. Of course, a thorough analysis of the data used by these indicators would be necessary to check the extent to which this really happens. For example, it might be the case that workers have not reported hazards that have been incorporated into their routine, and thus certain types of hazards would not be monitored by the frequency rate of near misses.

The analysis according to the sub-criterion 3.1 also provided insights about process safety monitoring. This task was performed as part of quality control, such as conducting tests on the performance of materials, and visual inspections for checking the maximum loads stored on a floor. Those procedures were part of a certified quality management system that was independent from the safety management system, which was focused on personal safety. Thus, safety staff and workers were not involved in the monitoring of process safety hazards, as they were not aware of the safety implications of the quality management procedures.

Concerning process safety, it is also worth noting that neither of the companies had an indicator to monitor accidents with material damages. Nevertheless, small accidents of this nature seemed to be frequent in both companies. For example, in one of the visits to a site of company B, the researchers noticed that a wall had collapsed, as a result of strong winds on the previous night. However, the safety specialist reported that he was not concerned with documenting and investigating that particular accident. The specialist took it for granted that the investigation of those types of accidents required technical knowledge in civil engineering. However, even the civil engineers who were legally responsible for the construction site were reluctant to take responsibility for process safety issues, since they, in turn, were relying on the knowledge of outsourced experts, such as those responsible for designing scaffolds, trenches and excavations. This context means that no one who worked full time on the construction sites was fully aware of all process safety hazards and on how they should be monitored. The possibility of monitoring safety performance during the whole product life-cycle was also neglected by both SPMS (sub-criterion 3.2). This could be done, for example, by monitoring safety during the product design stage, assessing the extent to which each design discipline (for example, architecture, utilities and so on) is complying with good practices of safety in design.

### *The SPMS Should Get Close to Real-time Monitoring*

The data processing and analysis cycles of the SPMS were relatively long. Only two out of the ten indicators adopted by the two companies were monitored on a daily basis: the index of training, and index of compliance and commitment. Overall, the evaluation based on sub-criterion 4.1 pointed out that feedback lagged substantially behind the moment at which events of interest took place. This feedback lag was due to: (a) the collection of data from past events (for example, accidents); (b) the collection of data concerned with unsafe conditions that could be in place for quite a while (for example, the lack of guardrails); (c) the fact that only one employee in each construction site, the safety specialist, centralized data collection, analysis and feedback, implying an overload (this was in conflict with sub-criterion 4.2); and (d) the delay involved in preparing reports and submitting them to the interested parties.

*The SPMS Should Take into Account Performance in Other Business Dimensions*

The SPMS of both companies did not have mechanisms for assessing how safety was performing in comparison with other areas (sub-criterion 5.1). Nevertheless, the existing SPMS had information which allowed some insights in this respect. For example, the SPMS could give visibility to the trade-off between safety and production, by calculating the ratio between the indicators PSAc (related to completion of safety activities) and PPAC (related to the completion of production activities). Concerning sub-criterion 5.2, companies A and B did not interpret the indicators from other areas from a safety perspective. Of course, some indicators did not have relevant links with safety (for example, number of complaints from clients) and interpretations from this perspective would be counter-productive. By contrast, other indicators were clearly relevant for safety. For example, both companies had indicators that monitored deviations from time and cost targets. If project time and cost are higher than expected, this should work as a warning that competition for the available resources (for example, money, time and labour) is increasing, and that resources that otherwise would be allocated to safety may be allocated elsewhere.

*The SPMS Should Balance the Trade-off Between Completeness and Ease of Use*

The trade-off between completeness and ease of use was managed intuitively by the safety staff of both companies, who were responsible for designing the SPMS. Since the safety staff were, at the same time, the designers and the main users of the SPMS, their goals and knowledge guided them in this regard. Therefore, they chose indicators that were fairly easy to collect and that provided meaningful information, from their viewpoint. This can be interpreted as an adaptive strategy adopted by the safety staff, in order to keep the SPMS compatible with the existing human, technical and financial resources (sub-criterion 6.1). Nevertheless, the centralization of tasks on the safety staff contributed to disseminate the idea that safety management was a subject of interest just for safety experts.

The only indicator whose objectives were misunderstood by managers (sub-criterion 6.2) was the PSAc. Instead of assessing whether the production activities were being carried out safely, as believed by management, the PSAc assessed if the activities of installing physical protections had been completed. While the ease of use was as a natural concern for the designers of the SPMS, the same concern did not exist with completeness. Indeed, assessing the completeness of a SPMS is much more difficult than assessing its ease of use, since it is impossible to know, and as a result to monitor, all hazards a complex system is exposed to. Nevertheless, the evaluation of sub-criteria 3.1 and 3.2 (the SPMS should monitor hazards throughout the socio-technical system) pointed out gaps concerning the completeness of the SPMS of both companies. The incompleteness of the SPMS was also a result of their over-reliance on the quantitative data produced by the indicators (sub-criterion 6.3). The frequency rate of near misses was an example of an indicator that provided quantitative data that was of little relevance in comparison with the qualitative data which was necessary for calculating the indicator (that is, the descriptions of near misses). Thus, the available pieces of evidence suggest that, in both companies, the trade-off between completeness and ease of use was pending in favour of ease of use.

## Conclusions

This chapter presented a set of criteria for assessing SPMS that is founded on the RE paradigm. Based on the RE perspective, it was possible to obtain insights that would not have been obtained based on criteria normally used to assess performance measurement systems in general. Due to the abstract character of the criteria, their application requires personnel who are familiar with RE and are domain experts. Moreover, the criteria, sub-criteria and sources of evidence need to be refined by extending its use to a large number of projects. As an indication of the usefulness of the criteria, a number of improvement opportunities were detected in the SPMS of two construction companies in which they were applied, such as: (a) the reports on safety performance should include key indicators from other areas, such as those related to project time and cost, as they can be proxy measures of the intensity of production pressures; (b) specific indicators could be designed to assess the trade-off between safety and production, such as the ratio between PSAc and PPAC; and (c) the SPMS should take a broader view on what counts as a hazard. Process safety and organizational hazards are neglected in favour of monitoring the more visible hazards emphasized by regulations (for example, falls, shocks and so on).

## Commentary

Performance measurements and performance indicators are the necessary foundation for management – of processes, of production, of safety and of resilience. A performance measurement system must furthermore be cost-effective. The notions of cost and effectiveness require a practical foundation. This chapter argues that the concepts of resilience engineering can be used for that purpose and illustrates that by looking closer at how two construction companies improved their safety performance measuring systems.

# Chapter 6
# A Framework for Learning from Adaptive Performance

Amy Rankin, Jonas Lundberg and Rogier Woltjer

There is often a difference between the way work is expected to be carried out (work as imagined) and the way it is actually carried out (work as done) in order to cope with complexity in high-risk work (for example, Hoffman and Woods, 2011; Hollnagel, 2012; Loukopoulos, Dismukes and Barshi, 2009). Events and demands do not always fit the preconceived plans and textbook examples, leaving nurses, air traffic controllers, fire chiefs and control room operators to 'fill in the gaps'. Examining practitioners' work shows a story of people coping with complexities by continuously adapting their performance, often dealing successfully with disturbances and unexpected events. Such stories can provide important information for organisations to identify system resilience and brittleness.

The literature on studies of people at work is replete with examples of sharp-end personnel adapting their performance to complete tasks in an efficient and safe way (Cook and Woods, 1996; Cook, Render and Woods, 2000; Koopman and Hoffman, 2003; Nemeth et al., 2007; Woods and Dekker, 2000). However, compensating for the system design flaws through minor adaptations come at a cost of increased vulnerability as systems are not fully tractable and outcomes are difficult to predict (Cook et al., 2000; Hollnagel, 2008; Woods, 1993). The terms sharp end and blunt end are often used to describe the difference between the context of a particular activity where work is carried out and factors that shape the context (Reason, 1997; Woods et al., 1994). The values and goals affecting sharp-end and blunt-end adaptive performance can be understood in terms of trade-offs such as optimality-brittleness, efficiency-thoroughness and acute-chronic (Hoffman and Woods, 2011; Hollnagel, 2009). Based on a number of overarching values and goals set by the blunt-end concerning effectiveness, efficiency, economy and safety, the sharp-end adapts its work accordingly. Effects of managerial level decisions, such as updating or replacing technical systems and procedures, are not always easy to predict and may lead to unintended complexity that affect the sharp-end performance negatively (Cook et al., 2000; Cook and Woods, 1996; Woods and Dekker, 2000; Woods, 1993).

Adaptations at the sharp-end have also been described as representing strategies (Furniss et al., 2011; Kontogiannis, 1999; Mumaw et al., 2000; Mumaw, Sarter and Wickens, 2001; Patterson et al., 2004). Strategies are adaptations used by individuals to detect, interpret or respond to variation and may include, for example, informal solutions to minimise loss of information during hand-offs or compensating for limitations in the existing human-machine interface (Mumaw et al., 2000; Patterson et al., 2004). Adaptations have also been characterised in terms of resilience characteristics, based on analyses of activities in terms of sense-making and control loops (Lundberg, Törnqvist and Nadjm-Tehrani, 2012).

Over time adaptations can have a significant effect on the overall organisation (for example, Cook and Rasmussen, 2005; Hollnagel, 2012; Kontogiannis 2009). Each individual decision to adapt may be locally rational, but the effects on the greater system may not have been predicted and far from what was intended. Although vulnerabilities may have been exposed they may not have been recognised as such when organisations do not know what to look for. Rasmussen (1996) describes this migrating effect in terms of forces, such as cost and effectiveness, which systematically push work performance toward and possibly over the boundaries of what is considered to be acceptable performance from a safety perspective.

In this chapter a framework to analyse sharp-end adaptations in complex work settings is presented. We argue that systematic identification and analysis of adaptations in work situations can be a critical tool to unravel important elements of system resilience and brittleness. The framework should be seen as a tool to be integrated into safety management and other managerial processes as it fills a terminology gap. The framework has been developed based on an analysis of a collection of situations where sharp-end practitioners have adapted (resulting in both successes and incidents), across domains that include healthcare, transportation, power plants and emergency services (Rankin, Lundberg, Woltjer, Rollenhagen and Hollnagel, 2013). The framework supports retrospective, real-time and proactive safety management activities. Retrospective analysis is done by using the method for analysis in the aftermath of events where an incident can be seen as a by-product of otherwise successful adaptive performance. The framework can further be used protectively to collect examples and identify patterns to monitor system abilities, predict future trends and appreciate 'weak signals' indicating potential system brittleness. In this chapter the framework is extended with a model of the interplay between framework categories using a control loop, which is demonstrated using two examples, one in crisis management and one in a health care setting. The illustration of the examples show how analysis of adaptations provides a means to improve system monitoring and thereby increase system learning. The framework has been applied as a learning tool but could in the future provide structure for teams and managers to 'take a step back' during real-time performance and examine if their strategies are having the desired effect, similar to what Watts-Perotti and Woods (2007) describe as broadening opportunities by blending fresh perspectives, revising assessments and replanning approaches.

## A Framework to Analyse Adaptations in a Complex Work Setting

The strategies framework describes the context in which an adaptation takes place, enablers for the adaptation and the potential adaptation reverberations in the system. An adaptation is in this framework referred to as a strategy. The context is described through identification of the current system conditions, the individual and team goals, as well as the overarching organisational objectives and forces which may play a role in the response. Resources and other enabling factors define the conditions necessary for implementation of the strategy. The potential reverberations of the strategy on the system are captured through an analysis of system resilience abilities (Hollnagel, 2009) and the sharp and blunt end interactions. Table 6.1 provides an overview of these framework categories and Figure 6.1 demonstrates how the categories relate in a dynamic environment. The overview is followed by two examples of how the framework can be used to analyse adaptations in everyday operations.

**Table 6.1 Framework categories and description**

| Category | Description |
|---|---|
| Strategy | Adaptations to cope with a dynamic environment. Strategies may be developed and implemented locally (sharp-end) or as part of an instruction or procedure enforced by the organisation (blunt-end), or both. To examine the adaptation effect on the system, strategy opportunities and vulnerabilities are identified. |
| Context | The factors that influence the system's need to adapt, e.g., events (disturbances), current demands. Feedback from the context provides input for the controller to assess the situation and prepare responses. Feedback may be selective or incomplete depending on system design and context. |
| Forces and objectives | Information on the manifestation of organisational pressures in a particular context. Forces are pressures from the organisation (e.g., profit, production) that may affect intentions and performance of the adaptation. Objectives are the organisation's overarching goals. |
| Resources and enabling conditions | Enablers for implementation of a particular strategy. Conditions may be 'hard' (e.g., availability of a tool) and 'soft' (e.g., availability of knowledge). This category extends the analysis of context in that it focuses on what allows (or hinders) the strategy to be carried out. This analysis can be further used to investigate information on the systems flexibility. |

**Table 6.1** *Concluded*

| Category | Description |
|---|---|
| Strategy goal | The identification of what the strategy is aimed at achieving. This can also be viewed as an outcome that the behaviour is aimed at avoiding. |
| Resilience abilities | Includes the four cornerstones; anticipating, monitoring, responding and learning, as described by Hollnagel (2009). The categories help identify a pattern of system abilities (and inabilities) with reference to the type of disturbances faced. |
| Sharp and blunt-end interactions | Recognition and acknowledgement of the strategy in different parts of the distributed system. A system must monitor how changes affect work at all levels of an organization, i.e., a learning system will have well-functioning sharp-end-blunt-end interactions. |

In Figure 6.1 a model is presented, demonstrating the interplay between the categories that constitute the framework. Building on the principle of a basic control loop (Hollnagel and Woods, 2005; Lundberg et al., 2012) we aim to recognise the dynamics of a socio-technical environment. The loop can be used to illustrate processes at different layers, such as the sharp- and blunt-end, of the system. Sharp- and blunt-end relations should be seen as relative rather than absolute as every blunt-'end' is the sharp-'end' of something else.

System variability may originate in external events such as changes or disturbances in the working environment or in natural variations in human and technical system performance. The context is shaped by the forces, objectives and demands on the system and may be affected by disturbances (no. 1, Figure 6.1). Monitoring of system feedback is required to assess what actions to take, and if there is a need to adapt (no. 2, Figure 6.1). A situation assessment is made based on the feedback provided by the system, organisational forces and objectives, strategy goals, identification of an appropriate strategy and resources available to implement the strategy (no. 3, Figure 6.1). The inner loop illustrates how the combined factors play a role in determining an appropriate action and that several options may be assessed prior to taking action. The trade-offs made can be described in terms of anticipating the opportunities and vulnerabilities that the adaptation may create. The process of recognising and preparing for challenges ahead has previously been described as the future-oriented aspect of sensemaking, or anticipatory thinking (Klein, Snowden and Pin, 2010). Attention is directed at monitoring certain cues, and responses are based on what is possible within the given context. Situation assessment and identification of possible responses is thus part of the same process and based on the current understanding (Klein, Snowden, Pin, 2010) (no. 3, Figure 6.1). The adaptation (no. 4, Figure 6.1) leads to changes in the environment which will affect the context. (no. 5, Figure 6.1).

The inner loop process is not necessarily explicit or available for full evaluation, particularly not in hindsight – a difficulty

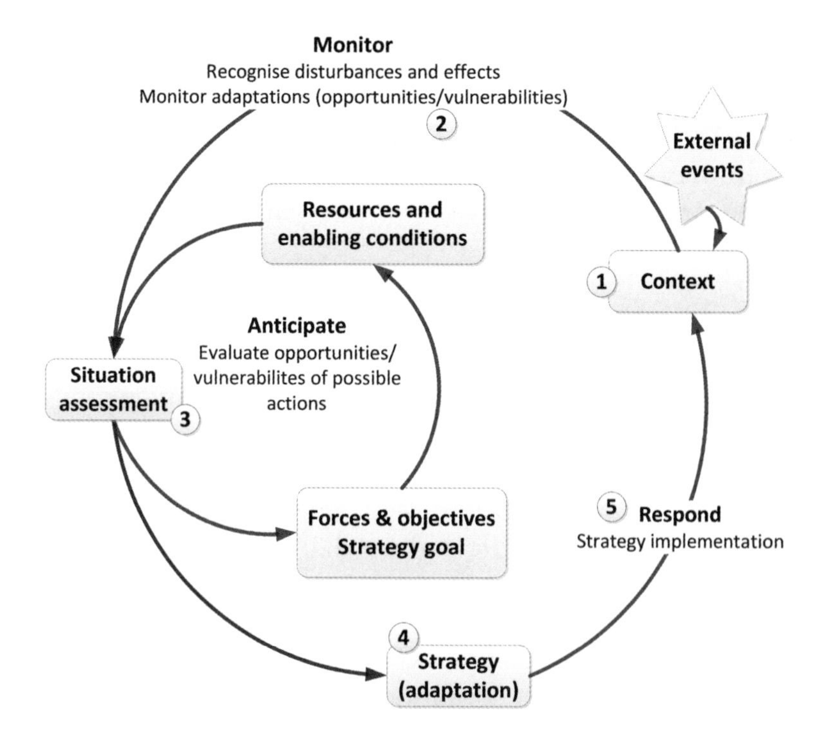


**Figure 6.1 A model to describe the interplay between the framework categories**

commonly referred to as hindsight bias (Fischoff, 1975; Woods et al., 2010). The analyst should therefore be cautious and avoid such biases when identifying and analysing the input in the inner loop to examine what shapes the situation and the selection of response (retrospective analysis). Trends and patterns of these factors can provide important information on what conditions and factors should be monitored to assess system changes in future events (proactive analysis). Balancing opportunities and vulnerabilities is not done on complete information at the sharp-end or the blunt-end and for sharp-end adaptations the time for interpretation is often restricted. Decisions and actions taken are based on the limited knowledge and resources available in a specific situation, that is, they are 'locally rational' (Simon, 1969; Woods et al., 2010).

## Demonstrating the Framework – Two Cases

Two cases are used to demonstrate how the framework can be applied to analyse situations where the system has to adapt outside formal procedure to cope with current demands. The cases are further visualised using the control loop model. The analysis focus is on how systems may learn from the analysis of adaptive performance and why monitoring everyday adaptations is important. In the first case a crisis command team adapts to a disturbance by re-organising the team, but fails to monitor the effects of the adaptation which subsequently contributes to several system vulnerabilities. The second case describes a situation where a team at a maternity ward successfully adapt to cope with high workload. The adaptation is acknowledged at managerial levels demonstrating how a system can learn from a sharp-end adaptation. A summary of the framework analysis can be found in Table 6.2 following the case descriptions and analysis.

### *Case 1 Reduced Crisis Command Team*

In this example a crisis management team is forced to adapt to the loss of important functions. The example comes from a Swedish Response Team simulation (Lundberg and Rankin, 2014; Rankin,

Dahlbäck and Lundberg, 2013) based on a real event: the 2007 California Wildfires. The team's overall aim is to support the Swedish population of 20,000 in the affected area by providing information and managing evacuations. One of the team's main tasks is to collect and distribute information regarding hazardous smoke in the area. This includes information on the severity of the smoke, required protection and where to get help if required.

Following a weather disturbance the command team is unexpectedly reduced from 18 to 11 members. The team responds rapidly by restructuring the team's functions and roles (no. 1, Figure 6.2). The disturbance and its effects were assessed by the team, led by the chief commander (no. 2, Figure 6.2). Adaptation to the new situation was enabled by the team's flexibility, which is part of the organisation design. To manage all key functions several people took on multiple roles. To assist the team in taking on roles outside their field of competence they used role descriptions found in the organisation's procedures (no. 3, Figure 6.2). The team was subsequently restructured, with several team members taking on multiple roles as well as roles outside their field of competence (no. 4, Figure 6.2).

The reverberations of the adaptation in this situation are depicted in Figure 6.3. Initially, the system responded efficiently to the initial disturbance and adapted by re-organising the command team structure to cover all necessary functions (no. 1, Figure 6.3). However, the organisation was now functioning in a fundamentally different way, creating new system vulnerabilities through an inefficient organisational structure with unclear responsibilities. The example demonstrates that adapting the team according to procedure does not necessarily mean that procedures are resources for action and that applying procedures successfully requires expertise and of how to adapt it to local circumstances (Dekker, 2003). The example demonstrates that adapting the team according to procedure does not necessarily mean that procedures are resources for action and that applying procedures successfully requires expertise and of how to adapt it to local circumstances (Dekker, 2003). The vulnerabilities of this situation were not detected by the team, or at least not explicitly acknowledged, and critical

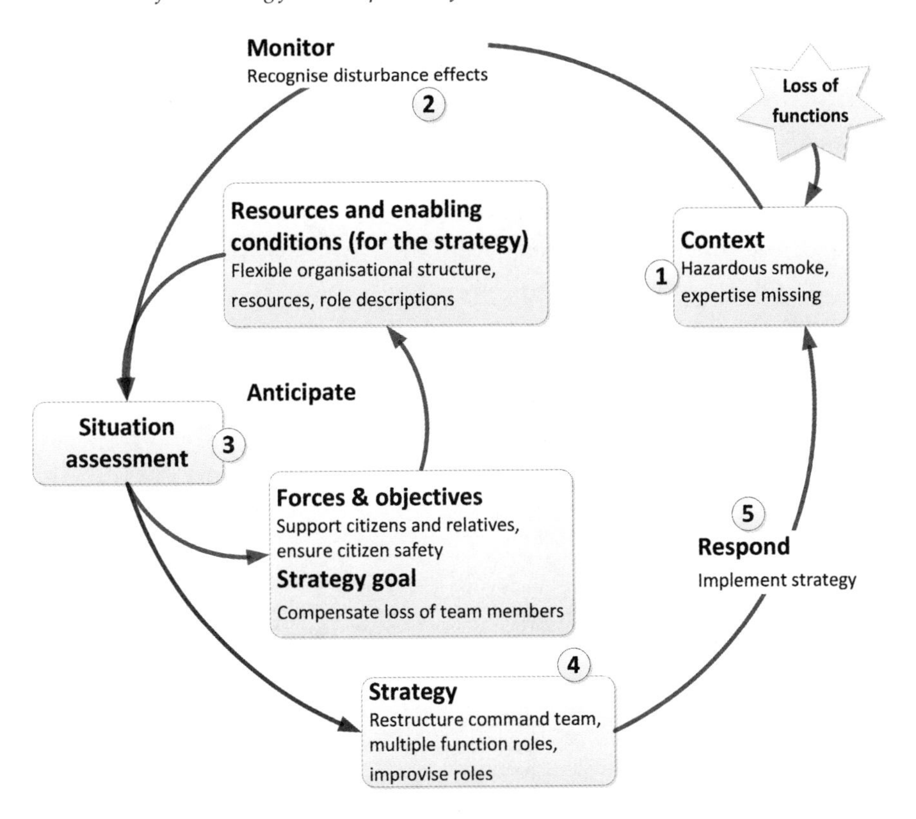


**Figure 6.2 Case 1 – Initial disturbance forcing the command team adapt**

information was lost, i.e., as the system adapts there is a failure to monitor how re-organisation affected the ability to adequately carry out tasks (no. 2, Figure 6.3). The difficulties in successfully fulfilling tasks are not completely undetected as joint briefings are held and incoming information is questioned (no.'s 3 and 4, Figure 6.3). However, strategies aimed at ensuring that tasks are carried out adequately are not sufficient to compensate for and unravel misinterpretations. Several critical aspects such as unclear responsibilities are overlooked and conflicting information within the team goes undetected (no. 5, Figure 6.3) (Lundberg and Rankin, 2014, Rankin, Dahlbäck and Lundberg, 2013). Although a procedure was in place, the team lacked the ability to adapt it to the local circumstances.

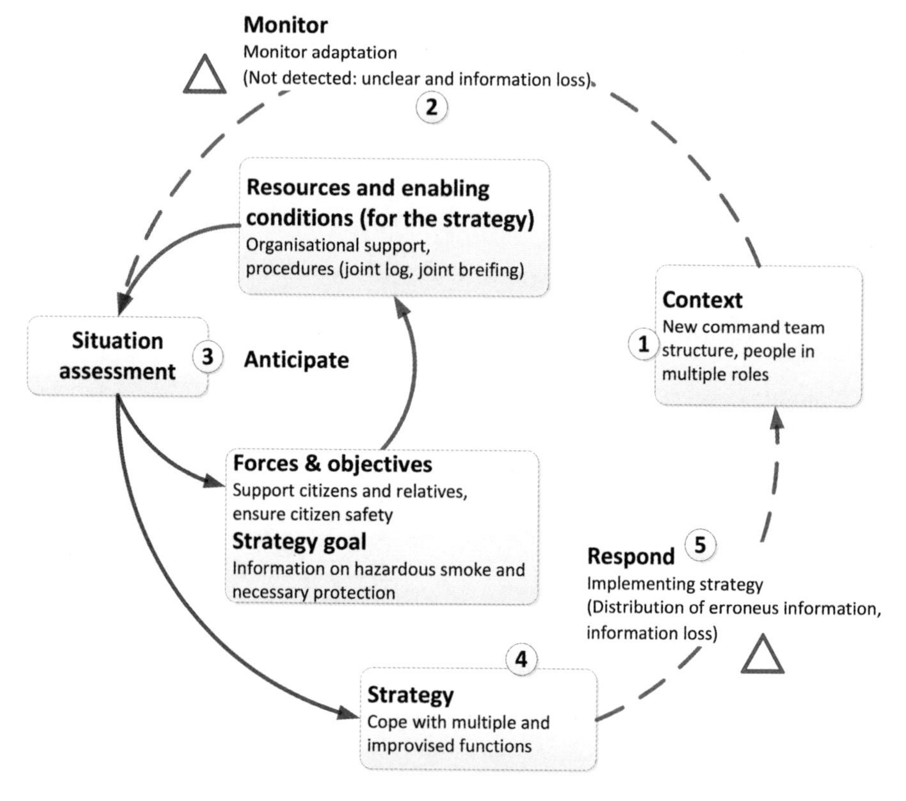


**Figure 6.3** **Failure to fully adapt and cope with the new situation**

*Case 2 High Workload at the Maternity Ward*

A remarkably large number of births one evening led to chaos at the maternity ward. The ward was understaffed and no beds were available for more patients arriving. Also, patients from the emergency room with gynaecological needs were being directed to the maternity ward as the emergency room was overloaded. To cope with the situation one of the doctors decided to send all fathers home of the newborn babies home. Although not a popular decision among the patients this reorganisation freed up beds, allowing the staff to increase their capacity and successfully manage all the patients and births. After this incident an analysis of the situation was performed that resulted in a new procedure for 'extreme load at maternity hospital'.

The system demonstrated several important abilities contributing to system resilience as it uses its adaptive capacity to respond to and learn from the event, which has been illustrated in Figures 6.4 and 6.5. Initially a large amount of patients are streaming into the ward, changing the contextual factors (no. 1, Figure 6.4). Based on an assessment of the current system status, the objectives and goals and the resources available, the doctor in charge decides to reorganise the resources to increase the capabilities necessary (no.'s 2, 3 and 4, Figure 6.4).

The strategy is realised and as a result the system has enough resources to cope with the high workload (no. 3, Figure 6.4, no. 1 Figure 6.5). Following the incident the managerial levels of the organisation detect and assess the occurrence, identifying system brittleness during high workload (no. 2, Figure 6.5). Based on the success of the sharp-end adaptation the blunt-end introduces a new procedure for managing similar situations (no. 3, Figure 6.5). The system thus demonstrates the ability to learn through well-functioning sharp- and blunt-end interactions (no. 4, Figure 6.5).

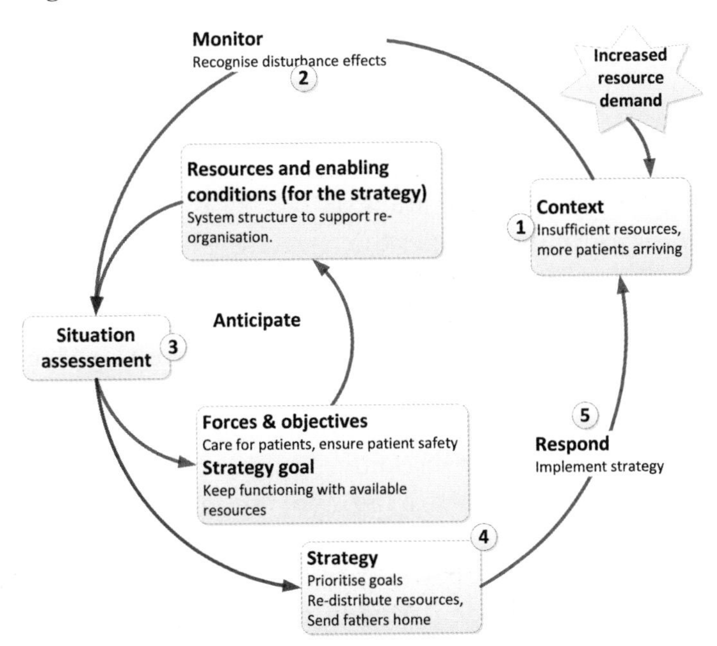


**Figure 6.4** **Case 2 – High workload at maternity hospital requires reorganisation**

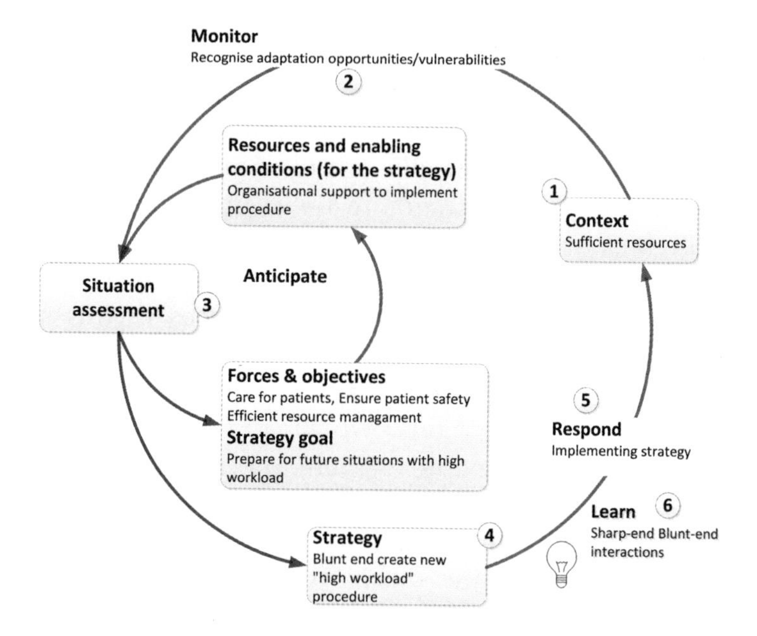


**Figure 6.5** **Case 2 – System learning through system monitoring and introduction of new procedure**

## Summary

The table overleaf summarises the framework analysis of the two cases described previously (Table 6.2).

Adaptations may change the system's basic conditions in unexpected and undetected ways, as shown in the first case. This phenomenon has previously been identified in studies of sharp-end adaptation as new technology is introduced leading to unforeseen changes in the work environment (Cook et al. 2000, Koopman and Hoffman 2003, Nemeth, Cook and Woods 2004). Hence, adaptations must be analysed not only in terms of their effects on the system, but also in terms of the new system conditions and the vulnerabilities (and opportunities) the changes may have generated. Although sharp-end adaptations are necessary to cope with changing demands in complex systems, they may also 'hide' system vulnerabilities by successfully adapting their work to increase efficiency and safety of the system. Recognising these shifts in work and what creates them provides necessary information to understand the operational environment and manage safety in a proactive way. An example of this is given in the second case where a sharp-end adaptation is acknowledged, recognised at managerial levels and used as a guide to improve the overall organisation.

**Table 6.2 Summary of framework analysis**

| Category | Case 1 Reduced Crisis Command Team | Case 2 High Workload at the Maternity Ward |
|---|---|---|
| Strategy | Restructure command team by taking on roles outside one's field of competence. Multiple function roles. Vulnerabilities include critical tasks not carried out, lack of domain knowledge, and lessened clarity of organisation structure. | Prioritise beds for patients and mothers and send fathers home (sharp end). Create new 'high workload' procedure (blunt end). |
| Context | Forest fires causing hazardous smoke. Citizens in need of support. Loss of important functions. | Inadequate resources due to many births. Patient sent from emergency room because overloaded. |
| Forces and objectives | Ensure citizen safety. Provide information to worried relatives. | Provide care and ensure patient safety. Efficient resource management. |
| Resources and enabling conditions | Flexible organisational structure. Resources. Role descriptions. | System structure to support reorganisation. Enough resources. |
| Strategy goal | Compensate loss of team members. | Manage workload with current resources' |
| Resilience abilities | Anticipating/Responding. | Responding/Learning. |
| Sharp- and blunt-end interactions | Blunt-end strategy enforced at sharp-end. | Sharp-end strategy turns into blunt-end procedure. |

*Monitoring Adaptations and their Enabling Factors*

It is not only when adaptations critically fail (for example, incidents and accidents) that they are of interest to examine. All adaptations outside expected performance provide information about the system's current state, its adaptive capacity and potential brittleness. Successful adaptations, failures to adapt or adapting to failure (Dekker, 2003) all provide key pieces of information to unravel the complexities of everyday work in dynamic environments.

By capturing everyday work and learning from adaptations it is possible to recognise what combinations of contextual factors, forces and goals create system brittleness as the system is forced to adapt. Similarly, the identification of adaptations allows insights into the individuals, teams and organisation's adaptive capacities (resilience), and provides important information for system design. It also enables successful work methods to be incorporated into the system through design and procedures, ensuring the enabling conditions are available.

The cases described above demonstrate the importance of monitoring the effects of adaptations over time. As systems are modified to cope with current demands the system may change in ways not foreseen or understood. Previous examples of adaptations using the framework demonstrate how reverberations of adaptations in one part of the system may cause brittleness in other system parts (Rankin et al., 2011; Rankin et al., 2014). An example of this is a hospital ward where the sharp-end strategy of ordering medication with different potency from different pharmaceutical companies was developed. The reason for this was that medicine packages of different potency from the same company looked almost exactly the same, creating an increased risk of using the wrong medication in stressful situations. Ordering different potencies from different companies resulted in packaging with different colouring, allowing the nurses to use colour codes to organise the medication. However, one time the medication at one company was out of stock and the order was automatically placed at a different company. The nurses were not informed of this, creating a situation where expectations regarding colour codes could lead serious incidents or even fatal accidents.

### *Integrating Resilience Analysis with Accident and Risk Analysis Methods*

The framework is intended be used to enhance current methods for learning from adverse events (accident and incident investigation) that build on defence-in-depth (Reason, 1997).

Accidents are traditionally analysed through identification of events, contextual factors and failed barriers at the sharp-end and the go upwards through the organisation toward blunt managerial ends, analysing failed barriers through organisational layers (defence-in-depth). The method presented in this chapter could add to this by incorporating forces and contextual factors that enable success through adaptation. The framework has previously been applied in several industries (Rankin et al., 2011, 2014) demonstrating the possibility of an applied tool for identifying system resilience and brittleness across industries.

The examples described in this chapter show the importance of monitoring and subsequently learning from adaptations. This requires monitoring in the *immediate, short* and *long* term. Assessments using the framework can identify how adaptations may affect other parts of the system and change the system conditions, creating new opportunities and vulnerabilities.

*Immediate monitoring* requires 'taking a step back' during real-time performance to observe the effect of the strategy and to make adjustments as required, i.e. broadening (Watts-Perotti and Woods, 2007). This would have been valuable in case 1 where the command team fell short as they did not identify vulnerabilities resulting from the re-organisation. Further development together with practitioners can allow the framework to be tested as a tool for 'broadening opportunities' real time (immediate) or as short term monitoring by practitioners in their work environment.

*Short term monitoring* requires well-functioning sharp-end – blunt-end interaction to identify and report adaptations for further investigation, as demonstrated in case 2. Learning how unplanned and undocumented adaptations have prevented a negative event and what enabled it may allow fast-track learning. Whereas the investigator by a traditional incident report only is provided with information on what went wrong (and perhaps an indication of why), the investigator after a report including otherwise successful adaptations may be provided with information on what has worked to prevent a negative situation, together with perceived factors of what enables the success. To put the framework to practical and short term use, incident reporting schemes can be adapted to include the analysis categories and perspective of the framework (such as the cases described in this chapter). Further directions for framework development is to us it as a guide for after action review following events where adaptations outside formal procedure are necessary. For incident reporting and after action review sessions, practitioners must be trained to identify not only incidents or situations with a negative outcome, but also report situations of resilience such as where they have invented or relied on an un-planned or undocumented strategy.

*Long term monitoring* to identify patterns and reverberations of every-day adaptations can be supported by the framework, and relies on short term monitoring to gather relevant data for analysis. Long term monitoring and analysis of situation in the aftermath of the event allows the identification of trends in adaptive performance and their reverberations over time which can serve as a guide to assess system resilience and prediction of how future changes may affect the overall system's abilities. However, the analyst must be cautious of hindsight bias, i.e. be careful not to assess the adaptation based on outcome, but examine what the circumstances, forces and enabling conditions are telling us about the system operations, its resilience and its brittleness. Similar to incident reporting, accident investigation can be enhanced and complemented by a resilience perspective using the framework as support. When an accident has occurred, information on how similar situations previously were successfully dealt with allows for better insight into 'work as done' rather than relying on 'work as imagined'. Furthermore, the framework can be used to enhance risk analyses by recognising strategies as part of 'work as done' and including an analysis of identified adaptation reverberations. Thus, vulnerabilities created as part of adaptations can potentially be foreseen and conditions to minimise the vulnerabilities can be strengthened.

The integration of resilience terminology into current safety management can provide insights into essential enablers for successful adaptive performance that may not surface through traditional reporting mechanisms. Analysing sharp-end work not only as it goes wrong but also what enables it to go right offers new perspectives on how to improve safety and the ability to deal with unexpected and unforeseen events.

## Commentary

In addition to the distinction between work-as-imagined and work-as-done, another important dichotomy is the difference between the sharp-end and the blunt-end. The two are furthermore not unrelated, since it often is the people at the blunt-end who prescribe how work should be done – but based on generalised knowledge rather than actual experience. Following the plea of Le Coze et al., this chapter presents a framework to analyse sharp-end adaptations in complex work settings. The analyses focus not only at what went wrong, but also looked at what actually happened: this focus on work-as-done brought forward experiences that would have been missed by traditional reporting approaches. By describing these using the four abilities (to respond, to monitor, to learn and to anticipate), the framework provides a window on how people deal successfully with unexpected and unforeseen events.

# Chapter 7
# Resilience Must Be Managed: a Proposal for a Safety Management Process that Includes a Resilience Approach

Akinori Komatsubara

The irregularity of socio-technical systems caused by various threats can lead to serious disturbance or accidents in our society. To avoid the disturbances, it is necessary to eliminate and diminish the threats or to establish barriers against the threats. That is the traditional safety approach. However, because these traditional safety approaches may not be sufficient to give stability to the socio-technical systems, a resilience approach may be needed. But in some cases that have been studied, and are described in this chapter, resilience has had no effects or resilience has caused specific accidents through functional resonance. Based on the case studies, a safety management process including a resilience approach is discussed and proposed as the conclusion.

## Introduction

Our modern society comprises various socio-technical systems. Socio-technical systems may vary in size and may vary in longevity – from temporary to permanent. When some irregularity happens in these systems, our society is significantly affected and disrupted.

Consider when a delay occurs in the schedule of a local train. We may not catch the flight that we had planned to take, and we may lose good business opportunities; that may have serious consequences for us. Our modern society is, after all, a fragile thing, like a glass sculpture made by various socio-technical systems. Therefore, specific management is needed to maintain the stability of the systems.

Factors that disturb the stability of the system are *threats*. All socio-technical systems are constantly faced with various threats. They can hardly escape from them. The kinds of threats can be classified into five categories:

a. Natural threats: these are natural disasters such as typhoons, earthquakes and heavy snowfall. Small animals or insects can also make serious threats, as seen in bird strikes on airplanes. Viruses and other infective agents can be a natural threat.
b. Social threats: these are pranks and malicious acts. Terrorism is the worst one. Children who put stones mischievously on railroad tracks are examples of social threats to railways. Recently, cyber-attacks have also become a serious threat.
c. Technical threats: these include equipment failure. It is well known that equipment failure is especially common when new technology is introduced. The troubles that occurred on B787 aircrafts in 2013 are one example. Old equipment often poses a threat as well, because however robust the equipment that was developed, it is impossible to escape from deterioration.
d. Service target threats: these are threats that arise when the demand that the socio-technical system serves outstrips supply. Large influxes of passengers on railways or patients at medical institutions can pose threats to the overall provision of reliable services of the socio-technical system.
e. Human threats: these are the so-called human errors and violations of the staffs. The decline in the level of the safety culture will accelerate the occurrence of human threats.

Specific countermeasures must be taken to prevent instability or accidents caused by these threats.

*Eliminating or Diminishing Threats*

Though this list is not necessarily complete, technical threats, human threats and service target threats are manageable. For technical threats, technical risk assessment and reliability engineering will be helpful. For service target threats, demand control will be effective. Medical triage for an emergency room and air traffic management (ATM) for flow control are the examples. For human threats, a traditional human error prevention approach is effective.

*Establishing Barriers Against Threats*

It is almost impossible to eliminate natural and social threats. Disaster prevention is therefore needed for natural threats. Sanitary management is as well. Security measures are needed for social threats. A firewall against cyber-attacks is one example.

These countermeasures are the traditional safety approach, and they are vital in keeping socio-technical systems steady and robust against threats. However, it may be impossible to escape every disturbance caused by these threats with the traditional safety approach. We can point out the following reasons.

- There are many occasions at the sharp-end where various threats are always emerging and disappearing without repeatable combination. An emergency room is a typical example. At such an occasion, because of technical, economical and social reasons, it would be impossible to make ideal and precise countermeasures for every threat actually.
- Unknown threats may exist that are beyond our imagination and anticipation.

Because of the above reasons, another countermeasure is therefore needed; we must take flexible countermeasures against the threats that emerge. That is the *Resilience Approach.*

There are two types of resilience approach. One is technical resilience and the other is human factors resilience. The typical example of technical resilience is seismically isolated structure for buildings. In this chapter, however, we focus on the human factors resilience. This is one aspect of human factors for safety; the other is traditional human error prevention. As Hollnagel (2012a) points Safety-I and Safety-II idea, the former will correspond to Safety-II and the latter traditional approach will correspond to Safety-I.

## Resilience Cannot Prevent Accidents or Cause Accidents

Though we expect safety to result from resilient behaviour of people involved, there may be some cases in which resilience has no effect in preventing accidents. Moreover, resilience may actually cause some accidents. The following are potential cases that should be considered.

### *Poor resilience Capability*

If the resilience capabilities of those who are expected to perform resilience behaviour at the sharp-end are poor, safety cannot be achieved. They cannot catch up with and solve the situation. Komatsubara, (2008a, 2011) points out four components required of an individual who is expected to perform with resilience.

1. Technical skills: it is obvious when we remember the Miracle of Hudson River in 2009. Without excellent technical skill to operate the aircraft that lost all power, the captain could not land in the Hudson River. If the technical skill of people involved is poor, slight disturbance of the system may easily impose an opportunistic or scrambled control mode of behaviour in COCOM (Contextual Control Model; Hollnagel, 1993), and it may easily lead to unwanted results. Therefore good technical skills including professional knowledge are the primary premise of resilience.
2. Non-technical skills: NTSs are indispensable for resilience. For example, to behave resiliently, we must anticipate and monitor the emergence of threats. Skills such as situational awareness and monitoring are, then, indispensable. Communication skills are also necessary to obtain good information for good responding.

3. Mental and physical health: mental and physical health are the basis for positive behaviour. Imagine when we have a cold. In that case, we cannot make good decisions. We may not have a positive attitude to make good resilience, either. Fatigue and healthcare management is therefore needed.
4. Attitude: vocational responsibility, social ethics, and a courageous mind are indispensable. In the Costa Concordia accident of the Italian coast in 2012, it is reported that the captain, who must behave resiliently to save passengers, escaped from the ship first. We cannot help thinking that his vocational sense of responsibility is insufficient.

Those who occupy a vocational position to achieve resilience should make self-efforts to enhance their resilience capability through these four components. At the same time, organizations must clarify the kinds and contents of capabilities that are needed for the staff members and support their development to enhance their resilience capability.

*Lack of Resilience Resources*

It is easy to understand that resilient safety cannot be performed without appropriate and sufficient resources.

In the Fukushima Nuclear Power Plant (NPP) accident in Japan in 2011, staff members of the NPP tried to perform resiliently, but resources were insufficient. They tried to recover the power supply of control panels, but emergency measures for the power supply had not been prepared. They therefore had to hastily gather car-batteries from cars parked in the plant yard and nearby car shops. This can be called a resilience behaviour, but this resilience to supply power could have been done more smoothly and effectively if emergency batteries as the resource had been prepared.

Organizations must anticipate and prepare the resources needed for the resilience by the staff members. Resources mean not only hardware such as tools and facilities, but also soft factors such as financial resources. The time allowed for resilience may also be important on some occasions.

*Missing Philosophy of Resilience*

All organizations have several purposes that they must achieve at the same time. The QCDS model of quality, cost, delivery and safety is a typical example. However, usually it is almost impossible to satisfy all the purpose at the same time. As the ETTO principle (Efficiency-Thoroughness Trade-off; Hollnagel, 2009) says, efficiency, that relates with cost and delivery, and thoroughness, that relates with quality and safety, often oppose each other. In that case, people tend to take efficiency rather than thoroughness. Moreover especially when an organization promotes cost-reduction, the people involved would tend to behave resiliently to achieve efficiency. The JCO criticality accident in Japan in 1999 is one example (Komatsubara, 2006). JCO was a small nuclear fuel processing company. Under a strong cost-reduction campaign of the company, the workers conceived the idea of producing liquid uranium fuel very efficiently, and worked resiliently to change the authorized production manual to pursue efficiency. A criticality accident then occurred, killing two workers. In this accident, resilient behaviour served not for safety but for efficiency. This means that if we regard resilience as an element of safety measures, we must establish and share the philosophy of safety before initiating resilient behaviours. Without a strong awareness of safety supported by good safety culture of the organization, resilience may start to drift to accidents.

*Possibility of Functional Resonance Type Accidents*

In situations where several people are involved, functional resonance type accidents (Hollnagel, 2012b) may occur when the combination of each resilient behaviour is inappropriate. The following are some such cases.

*Case1) When people involved do not have the same context*
This is the case which I encountered. This case is a private one, but if an accident had occurred, large disturbance might have been given to the road traffic as a socio-technical system.

The situation is shown in Figure 7.1. I was driving a car. In Japan, as in the UK, the driving lane is on the left. I intended to turn to the right at crossing A, and turned on my right turn signal and slowed down. At the same time, a public bus was coming from the opposite side, and the driver turned on the right turn signal, too. I understood that this bus would turn to the right at the same crossing A, and I started to turn to the right. This bus continued to come straight on, however, without slowing down, and we very nearly collided. I understood later that this bus had intended to turn right up ahead, at crossing B.

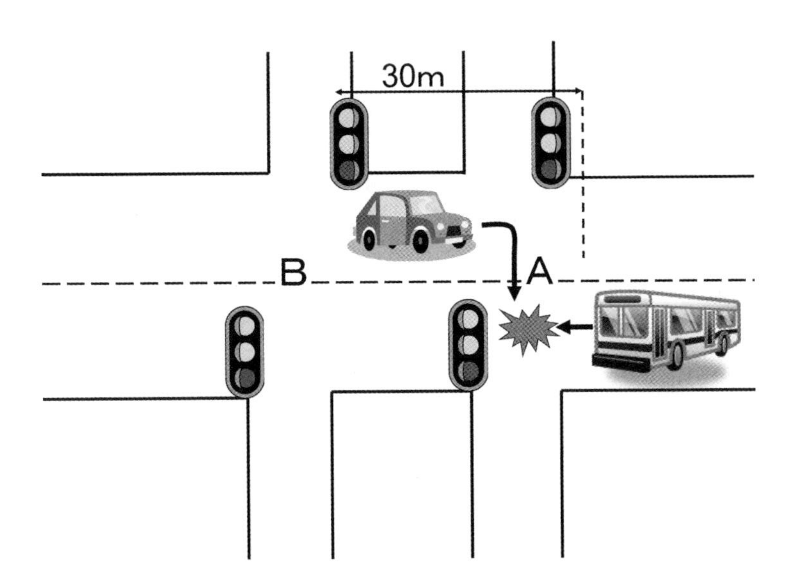


**Figure 7.1 The near-miss situation that I encountered**

The traffic laws of Japan determine that drivers who wish to turn right or left should switch on the turn signal 30 metres ahead of the crossing. The behaviour of the bus driver was therefore appropriate legally. Probably, as I slowed down before the crossing A, he might have imagined that I understood his intention and would stop until his bus passed through, and he therefore continued to come straight on. I thought that this bus would turn to the right at the very crossing A, however, because this bus switched on the right turn signal before crossing A.

Therefore, my behaviour of starting to turn to the right before the bus passed through the crossing was very natural. The FRAM analysis of this case is shown in Figure 7.2.

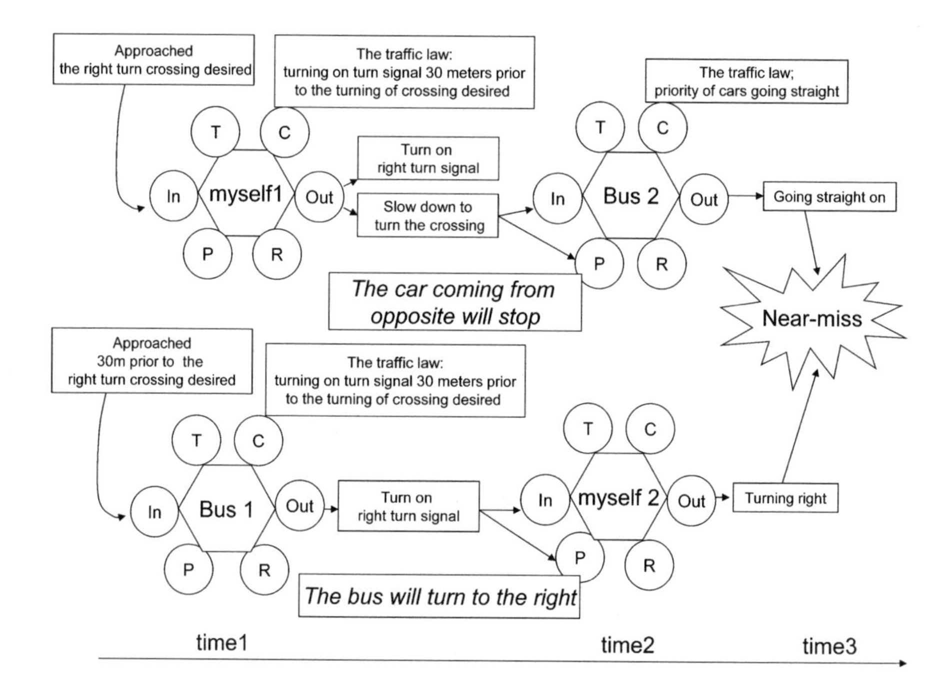


**Figure 7.2 The cars' near-miss case expressed by FRAM**

In this case, the bus driver and I behaved resiliently based on the different understanding we each had of our counterpart's intention; the combination of our preconditions was inappropriate. That is, doing resilient behaviours based on contexts or preconditions that differ from one another may lead to functional resonance accidents.

*Case2) When the people involved behave under different controls*

The near-miss incident of flights JAL907 and JAL958 in Japan in 2001 is a typical example (JTSB, 2002). In this incident, two aircrafts were coming from opposite directions at the same altitude. To avoid collision, one aircraft obeyed Traffic alert and Collision Avoidance System (TCAS) instruction and the other obeyed air traffic control (ATC) advice, and when both descended, a near-miss occurred. Figure 7.3 illustrates the simplified situation expressed with FRAM.

In this case, both captains had the same understanding that a collision would occur unless they changed their altitude or the course. However, each control was different even though both that had been based in resilient behaviour to avoid the collision.

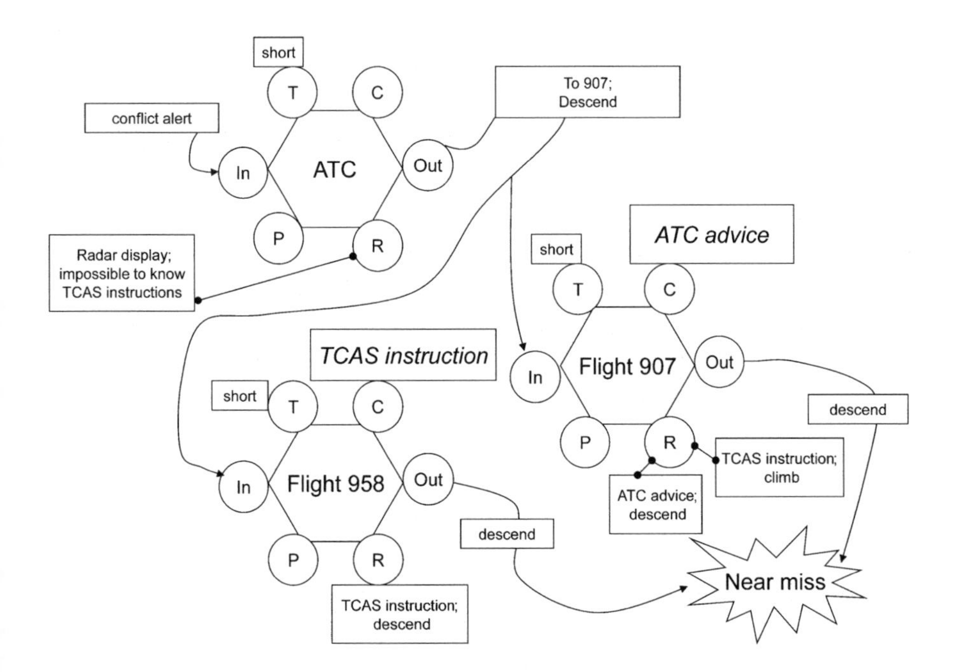


**Figure 7.3 The aircrafts' near-miss case expressed by FRAM**

A functional resonance accident may occur when several people involved have different preconditions, different controls or inappropriate timing, when performing resilience behaviour. To avoid this type of accident, then, we must consider how they can have same precondition, same control and appropriate timing to act with resilience.

## No Blame Culture is Needed for Resilience

Based on Komatsubara (2008b), the position of those who perform resilience can be classified into a 2 x 2 matrix.

One aspect is the position of resilience; the resilience is a vocational or civil one.

Vocational resilience is expected of the people who serve the public as their job. Medical doctors, airline pilots and fire fighters are examples of such jobs. These professionals are expected to have resilience naturally and essentially.

Civil resilience is expected of the people who accidentally encounter the situation that needs resilience. Hollnagel (2011) points out the four essential factors for resilience of responding, monitoring, anticipating and learning. Strictly speaking, therefore, civil resilience may not be the resilience in the sense of resilience engineering because it usually lacks in the factors of anticipating and monitoring, but this chapter includes it to deepen the discussion. Consider when we take a seat next to an elderly person on a train, and suddenly the elderly person suffers a heart attack and loses consciousness. What should we do? We could take the attitude of a bystander, and escape from the situation, but we would most likely do something resiliently to save the elderly person.

The other aspect is that the results of the resilience done by those involved can or cannot contribute to the direct damage as injury or death of the people themselves. In other words, the resilience is only for others or for the people including him/herself. In the latter case, they could not take the attitude of a bystander because they would result in his/her injury or death if he/she does not act resiliently.

Table 7.1 shows examples of these positions of resilience.

**Table 7.1 Classification of the position of Resilience Behaviour**

| | | Direct damage is imposed if they take the stance of a bystander? | |
|---|---|---|---|
| | | **Yes**; Direct damage is imposed on him/her if he/she does not conduct resilience. | **No**; Direct damage is not imposed if he/she does not conduct resilience. The resilience is basically for others' happiness. |
| Position of resilience | Vocational | Airline pilot | Medical doctor |
| | Civil | Passengers who are taking a bus when the bus driver suddenly loses consciousness. If the passengers do not do something, they themselves may be injured or killed. | People who have just encountered a situation that calls for someone's resilient help, but they can take the attitude of a bystander. |

Especially when they serve others only, a blameless culture is needed with both vocational and civil resilience. Resilient behaviour does not always lead to a good result. Slight human errors may be inevitable in resilience. Moreover, in civil resilience, they may have little technical skills to incorporate resilience. In those cases, however, the people involved do their best to incorporate resilience behaviour at the very situation at the very moment, unwanted results may occur. In this situation, if any blaming voice with hindsight arises, they, or we, may lose the motivation to incorporate resilience. This is because they, or we, do not want to take responsibility for the unwanted result. Therefore, a blameless culture based on the idea of the *Good Samaritan law*, based on Chapter 10, paragraphs 29–37 of the Gospel of Luke is required.

## Resilience Must Be Managed

Let's consider soccer games. Players play resiliently to win the goal. Victory depends on their efforts. In addition, we must notice that the head coaches have important role for the victory as well. Of course, they must respect players' decision and ensure flexibility for them because threats essentially occur at very sharp-end. They must, however, behave resiliently to be good at keeping the players in the game in hand. Moreover we must also notice that the coaches have already started their role prior to the game. They must investigate the opponent to make a strategy of the game, and develop training of each player to perform their best in the game. They may pay attention to the equipment like players' spiked sports shoes, too. It is no exaggeration to say that the performance of players in the game depends on the prior management by the coach.

It is completely the same in safety, too.

When the sharp-end is fighting resiliently with threats, the blunt-end must control the sharp-end resiliently, for example, to avoid functional resonance accidents. However it is also very important for the blunt-end to make a steady effort so as to let the sharp-end win the threats. This must be conducted before the sharp-end actually starts resilient behaviour.

After all, the blunt-end organizations must develop some daily management and preparation for resilience of the sharp-end.

## How Should We Manage?

Based on the discussion before, we should start the management from a traditional safety approach. At first, we must understand that resilience approach does not exist independently.

First of all, as far as possible, we must anticipate and identify threats that the sharp-end may encounter, and we must try to eliminate or diminish the threat or we must create barriers against that threat.

By doing so, we can avoid entangling sharp-end in a meaningless resilience. Consider medication dose error, for example. If we will not give the right medicine to the right patient, doctors must act resiliently to cure the patient for the medication dose error. The resilience, however, can be essentially avoided if the right dosing would be performed; that would be accomplished with traditional human factors approach.

However, we may not be able to establish the countermeasures that are enough to overcome the threats. Human errors would happen because 'to err is human'. In that case, resilience is needed. Moreover resilience is needed against the threats which are beyond or out of our forecasting and imagination. As for the threats which are out of imagination, it is impossible to anticipate, of course. But it is possible to determine the unwanted events for the system. For example, we know how serious the loss of all power supply of NPPs would be. Therefore, we could prepare resilience against occurrences of all power supply loss as one of the unwanted events.

After that, we should enhance the resilience capabilities of individuals at the sharp-end. To do so, we must identify the kinds of resilience capabilities that are needed for the individual, considering the kinds of expected threats and unwanted events. We must also monitor whether the sharp-end has satisfied the identified capability or not. Then, we should make and develop training programs. In addition, in some cases, we must prepare some resources needed for the resilience activity. Moreover, when several people are involved in the situation, some management shall be needed to avoid functional resonance accidents caused by mismatching of each resilient behaviour. As for the car near-miss case that I encountered, if a signboard that just says 'Beware of oncoming vehicles!' had been set up, it might have enough effect to coordinate my precondition to turn to the right. At the near-miss case of two aircrafts, it is now regulated that pilots should only respond to TCAS resolution advisory (RA). This regulation helps pilots avoid having different controls when TCAS issues.

No matter how carefully we prepare, however, some situations that are caused by unexpected threats beyond forecasting may occur. In that case, prepared countermeasures may not be effective, but we expect that the people involved will act resiliently to settle and to recover from the situation. But this might bring unwanted results. Even if the results are not favourable, we should not blame it from hindsight. This means that blunt-ends must cultivate society to understand no-blame culture.

We must understand that all activities above should have been performed and well prepared by the blunt-end before the sharp-end encounters the very situation that needs resilience. Moreover, after the sharp-end has encountered the situation that required resilience, the blunt-end must review and evaluate that those prior activities were enough. Through the review, they must make corrections of the prior activity if needed. Those reviews will bring learning for the management, too.

To sum up, this chapter proposes the safety management system (SMS) model including robust and resilience as Figure 7.4. In this study, PDCA model of plan, do, check and act, is taken because PDCA model is very common in organizational management. If we take PDCA model, we can expect that Figure 7.4 process will be smoothly accepted by the organizational managers.

By promoting and repeating this PDCA cycle by the blunt-end, we can expect to establish the robust but resilient safety of the system.

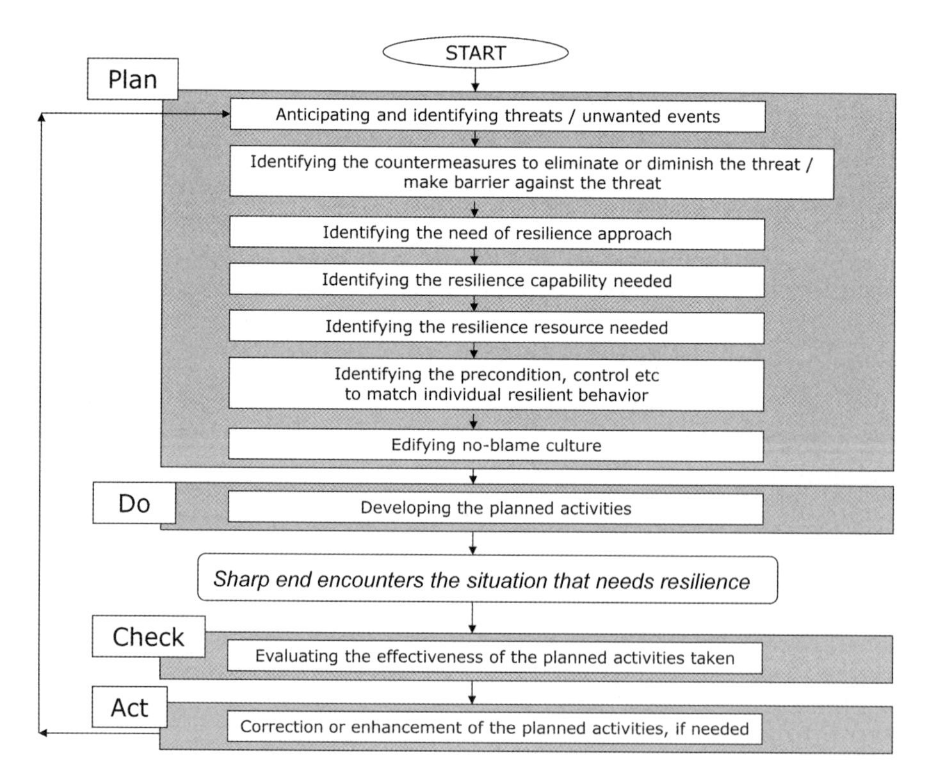


**Figure 7.4 Safety Management System Model including Robust and Resilience Approach**

## Conclusion

In this chapter, a safety management model including the resilience approach has been proposed to keep the stability, and to increase safety of, the socio-technical systems. The resilience approach does not exist independently from the traditional safety approach including human factors. Moreover, the resilient safety approach may result in failure without prior and good management. If no prior management had been made (even if the sharp-end succeeded in achieving safety) of course we can learn from it for the next resilience but we must say it was just lucky. To increase the safety of socio-technical systems with the resilience approach, we need to understand the position of the resilience approach in the whole strategy of safety.

Without prior management, resilience would not win.

## Commentary

This chapter puts forward the argument that we should be careful not to look at resilience – or resilience engineering – as a panacea. The resilience engineering perspective, and more specifically the Safety-II perspective, is not intended as and should not be used as a replacement for safety management. It is rather a complement to or a new but crucial facet of safety management. A resilience-based approach should not be relied on independently from a traditional safety approach – including a focus on human factors. To increase the safety of today's intractable socio-technical systems it is necessary to understand the position of the resilience approach within the whole strategy of safety.

# Chapter 8
# A Case Study of Challenges Facing the Design of Resilient Socio-technical Systems

Alexander Cedergren

## Introduction

In the general sense of the word, resilience refers to the ability to withstand and recover from various types of stresses. This concept has gained significant popularity in as disparate fields as psychology, engineering, and ecology (Birkland and Waterman, 2009; de Bruijne, Boin and van Eeten, 2010; Woods, Schenk and Allen, 2009). The emphasis in the field of resilience engineering lies at the abilities of individual teams and organisations to adapt and survive in the face of various disturbances. However, the resilience studied at one level depends on the influences from levels above and below (Woods, 2006). These different levels operate at different timescales, and there may be tensions and incompatibilities between them, which impacts the nature of the system's resilience (McDonald, 2006). For this reason, it is important to investigate the relationships between the various levels of a socio-technical system, in order to understand the ways in which resilient performance is achieved. This chapter therefore aims at paying attention to the way resilience of a socio-technical system is influenced by the interplay between various stakeholders in a multi-level and multi-actor context.

As a basis for the analysis outlined in this chapter, a case study of the decision-making process at the design stage of railway tunnel projects in Sweden is presented. Decision-making in this type of projects is undertaken in settings characterised by a multiplicity of diverse actors and perspectives, where no single actor has the authority to make a final decision. In this way, the present context is characterised by some crucial differences from studies of resilience in individual teams and organisations, and the chapter aims at contributing with new insights as well as areas for further consideration to the field of resilience engineering.

The case study presented in this chapter takes as a point of departure four factors described by Woods (2003) as important challenges for managing a system's resilience. These factors include:

- Failure to revise assessments as new evidence accumulates
- Breakdowns at the boundaries of organisational units
- Past success as a reason for confidence
- Fragmented problem-solving process that clouds the big picture

These four factors are used as a basis for the analysis presented in the subsequent sections, and in this respect, the chapter has some similarities with the study presented by Hale and Heijer (2006). Before the case study is further described, some background of the decision-making process in Swedish railway tunnel projects will be outlined.

## The Decision-making Process at the Design Stage of Railway Tunnels

The case study presented in this chapter builds upon semi-structured interviews with a total of 18 respondents involved in the decision-making process at the design stage of railway tunnel projects in Sweden. Six different railway tunnel projects were included in the study, and together these projects encompass 28 tunnels, ranging in length between 180m and 8.6km. Preliminary findings from this case study have been presented in Cedergren (2011), and for a more extensive presentation of the case study, see Cedergren (2013).

Two main sets of actors are involved in the studied decision-making process. These key players are shown in Figure 8.1, which schematically outlines the decision-making process in question. The first set of actors will be referred to as the project team. The project team consists of employees from the Transport Administration, which is the national authority responsible for building and maintaining railway infrastructure (as well as road infrastructure). In addition, the project team includes various consultants, which are appointed by the Transport Administration to this type of projects, particularly for conducting risk assessments and other types of safety-related documentation. The second set of actors will be referred to as the local actors, and consists of the local building committee and the local rescue service from the municipality in which the tunnel will be constructed. The role of these players will be further described in the following sections.

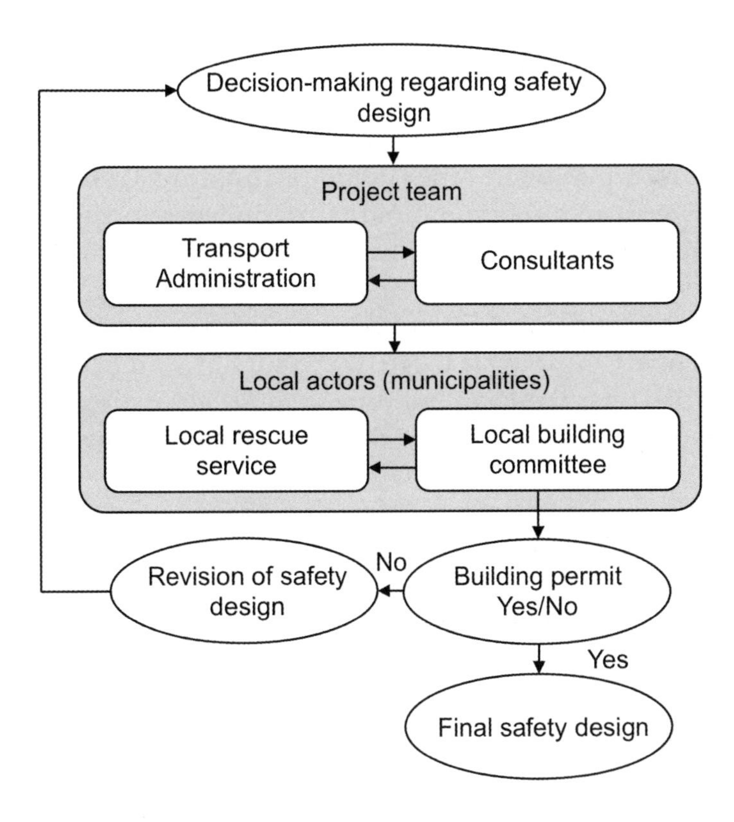


**Figure 8.1** **Schematic outline of the main actors involved in the decision-making process at the design stage of railway tunnel projects**

## Results and Analysis

As described above, the decision-making process in the studied railway tunnel projects has been analysed from a resilience engineering perspective, with special consideration of the ways in which the interplay between various stakeholders in a multi-actor and multi-level context influences the system's ability to perform resiliently. The results from this analysis are outlined in the following sections, under the four headings that were used as a point of departure.

### *Failure to Revise Assessments as New Evidence Accumulates*

One of the key issues for decision-making at the design stage of railway tunnels relates to the provision of means of evacuation from the tunnels. While the Building Codes regulate means of evacuation for most other types of buildings in Sweden, they are not applicable to railway tunnels. Instead, another set of laws and regulations are applied to this type of constructions. According to these laws, an approved building permit is required from the local building committee before a railway tunnel can be taken into operation. This means that the local building committee (the local authority) needs to approve the design suggested by the Transport Administration (the national authority). In the studied projects, the local building committees experienced that they did not have sufficient competence on questions related to risk, safety or means of evacuation. For this reason, they usually appointed the local rescue service as their expertise on these issues. As a result, these local actors held a prominent position in the decision-making process.

Another important aspect of the laws applied in railway tunnel projects is the lack of a risk acceptance criterion for the design of railway tunnels. For this reason, the Transport Administration has issued an internal handbook for the design of railway tunnels, which builds upon a risk-based approach. According to this handbook, railway tunnels need to be equipped with various types of safety measures in such way that the overall level of risk, as estimated in risk assessments, meets the established risk acceptance criterion. One of the most important types of safety measures in railway tunnels is the evacuation exits, which are located at regular intervals inside each tunnel. These evacuation exits are also used by the rescue service as intervention points in the case of fires or other types of emergencies.

Due to the significant costs associated with each additional evacuation exit, the Transport Administration are normally reluctant to design tunnels with a larger number of exits than deemed required from their risk assessments. However, in most of the projects included in this study, the distance estimated in these risk assessments was seen as too long by the rescue services. For this reason, the rescue services rejected the risk-based approach adopted by the Transport Administration. In contrast to the risk-based approach to decision-making adopted by the Transport Administration, the rescue services adopted a more deterministic approach. In their view, the occurrence of a fire inside a tunnel was taken as the starting point for decision-making, regardless the probability of such event.

In order to minimise the walking distance inside a potentially smoke-filled tunnel, the rescue services plead for a shorter distance between the evacuation exits. The different stakeholders thus drew on different types of evidence claims for proving their point in the decision-making process. The "evidence" highlighted by the rescue services rested upon experience from rescue operations in similar environments, whereas the "evidence" presented by the Transport Administration was based upon the outcome from risk assessments and cost-benefit considerations. These divergent perspectives implied that the problems in question were looked upon in contrasting ways, and different solutions were advocated. This resulted in a decision-making situation in which none of the players were able, or at least not willing, to revise their assessments when the other player presented new evidence.

### *Breakdowns at the Boundaries of Organisational Units*

The various types of evidence presented by the different players reflected their diverse framings of the risks associated with railway tunnels. As a result, controversies arose in several of the studied projects, particularly with regards to the distance between evacuation exits. As described above, the Transport Administration used a risk-based approach for estimating this distance. However, in most projects, the rescue services did not agree upon the distance between evacuation exits estimated by the Transport Administration. In order to intervene in the case of emergencies, they argued for a shorter distance between these evacuation exits. They also argued for a need to provide additional safety measures and rescue equipment, such as smoke ventilation and water pipe systems. Unless these additional measures were included in the design, the local authorities were not willing to approve the building permit. In this way, members of the project teams in several projects experienced that the rescue service tried to "kidnap" the building permit.

As a result of the additional demands raised by the rescue service in many projects, the Transport Administration experienced that they were trapped in a double bind. On the one hand, agreeing upon the additional demands raised by the rescue service would lead to increased costs for the project. On the other hand, rejecting the demands raised by the rescue service would imply that the project would be delayed, which also resulted in increased costs. Consequently, no matter what actions they took, it would lead to undesired outcomes.

In a similar way, the local actors experienced that they were trapped in another kind of double bind. On the one hand, approving the design suggested by the Transport Administration could potentially (in the aftermath of an accident with severe consequences) imply that they would be blamed for having approved the construction of a railway tunnel with an insufficient safety standard. On the other hand, disapproving the building permit would imply that they would be blamed for delaying the project, which typically constituted an important infrastructure investment for the region in question. In this way, no matter what actions they took, they experienced that they would be blamed for the outcome of their decision.

The double binds experienced by both of the main players involved in the decision-making process resulted in controversies and deadlocks in many projects. These breakdowns at the boundaries between the various organisations illustrate the challenges associated with decision-making in settings characterised by a multiplicity of stakeholders with diverse roles and perspectives.

*Past Success as a Reason for Confidence*

In order to reconcile the deadlocks emerging in several projects, and thus attaining an approved building permit, the Transport Administration agreed upon some of the demands raised by the rescue services. For example, in one of the projects the Transport Administration agreed upon the demands raised by the rescue service on providing a water pipe system. However, by agreeing upon this safety measure in this particular project, the rescue services in each succeeding project also raised demands on provision of the same type of water pipe system. In this way, each safety measure that was approved in a specific railway tunnel project gave rise to a "precedent", that is, a decision made in one project which was used as a justification for making the same decision in future projects. As a result, the level of safety measures was gradually raised in each consecutive project.

As a reaction to the increased demands for safety measures in each new project, several members of the project teams emphasised the need to take a cost-benefit perspective on safety investments in railway tunnels. In their view, the low probability of accidents in railway tunnels did not justify the large amount of safety measures demanded by the rescue services, and the lack of previous accidents in railway tunnels was taken as a reason for confidence in this viewpoint. For these players, the establishment of precedents created frustration. In addition to the increased costs and delays associated with the establishment of precedents, the large influence from decisions reached in previous railway tunnel projects downplayed the role of the risk assessments. In this way, the risk assessments were not the only, and not even the most important, basis for decision-making. Rather, decisions were in many cases highly influenced by negotiations between the main actors involved in the decision-making process. For this reason, members of the project teams questioned the value of spending considerable resources on conducting risk assessments when these assessments, in the end, were not used as a basis for decision-making.

So far, it can be concluded that an informal decision-making process was evolving in parallel to the formal decision-making process, in which risk assessments were conducted as a basis for decisions. As a result of this informal process, precedents, negotiations, and power relations between the different players were highly influential for reaching a final decision. The next section describes the ways in which the railway system's ability to perform resiliently was affected by these processes.

### *Fragmented Problem Solving Process that Clouds the Big Picture*

The controversies arising in many of the railway tunnel projects arose as a result of the diverse framings held by the local and national authorities. Due to the need for an approved building permit in these projects, the viewpoint adopted by the local actors gained significant influence in the decision-making process. This viewpoint primarily focused on railway tunnels from a local perspective, that is, the way each tunnel was equipped with different types of safety measures (in order to ensure evacuation and rescue operation). However, due to the large emphasis on local aspects, limited attention was devoted to the railway system's performance from a regional or national perspective. The bigger picture of the railway system's functioning, including its ability to perform resiliently in the face of failures, was thus overlooked. In this way, no single actor had a coherent view of the railway system and its functioning at the overall level. This was particularly clear with regards to the choice of tunnel type and provision of means of evacuation. Different ways of building railway tunnels are associated with different solutions for means of evacuation, and two of the solutions adopted in the studied projects are illustrated in Figure 8.2.

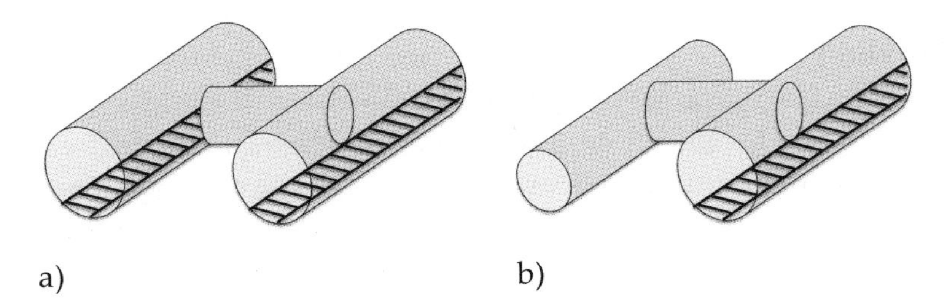


**Figure 8.2** **Different tunnel types with different solutions for providing means of evacuation. A) schematically illustrates a tunnel type with two parallel railway tunnels, and B) schematically illustrates a single-line tunnel type with a dedicated parallel evacuation tunnel (walking paths for evacuation not illustrated in the figures)**

Figure 8.2a shows a solution in which two parallel railway tunnels are connected with rescue tunnels at regular intervals. Figure 8.2b shows a single-line railway tunnel with a dedicated parallel evacuation tunnel. Due to technical constraints, the evacuation tunnel in this latter design had to be constructed of almost the same size as the full-sized rail tunnel. If the evacuation tunnel had been made somewhat larger, it would have been possible to prepare this tunnel for using it as a redundant railway tunnel (in effect, transforming it into the same type as shown in Figure 8.2a). According to several respondents, this solution would only imply a negligible additional cost, but provide the railway system with a significantly improved ability to continue operations in the face of failures affecting the single-line railway tunnel. However, since focus in the decision-making process was concentrated to the local features of each tunnel, consideration of the ways in which each tunnel affected the performance of the larger part of the railway system was restricted. In this way, the railway system's ability to perform resiliently from a regional and national perspective gained limited attention in the projects.

## Discussion

In this chapter, special consideration has been paid to the ways in which resilient performance of a socio-technical system is shaped by the interplay between actors with diverse roles and perspectives. As shown in the case study presented in the previous sections, the various players involved in this type of multi-actor and multi-level setting framed risk in different ways, and as a result, they drew on different evidence claims in order to influence the risk management process (cf. van Asselt and Renn, 2011). This decision-making situation, in which no single actor had superior authority, resulted in disagreements and controversies among the key players. Due to these disagreements, both of the two key players experienced that they were trapped in different types of double binds, that is, decision-making situations leading to undesired outcomes no matter what actions were taken (cf. Dekker, 2006). While the municipal actors experienced a double bind related to blame, the project team experienced that they were facing increased costs no matter what actions they took. In parallel to the formal decision-making process, in which risk assessments were conducted as a basis for decision-making, an informal process evolved. In this process, "precedents", that is, decisions made in previous projects, played an important role. As a result, the outcome of risk assessments was downplayed, whereas greater weight was placed on efforts to reach agreements through negotiations between the different players.

While there is no doubt that studies of individual teams and organisations provide valuable insights to the field of resilience engineering, this chapter demonstrated that it is important to also take cross-organisational aspects into account (see also Mendonça, 2008). In large-scale socio-technical systems, resilience is an ability characterising the system at the macro-level. However, the macro-level abilities are created by the individual actions and decisions taken at the micro-level (cf. Vaughan, 1996). For this reason, understanding of the way that resilience of socio-technical systems is fostered requires close consideration of the cross-scale interactions between various levels (see also McDonald, 2006; Woods, 2006), that is, the link between activities at the local level, and their effects on the global level. By restricting studies to a single actor's perspective, the ways that resilient performance of a system is affected by the interplay between various players in a horizontal as well as vertical dimension go unheeded. For this reason, the objective of this chapter was to apply a resilience engineering perspective to the analysis of a large-scale socio-technical system by adopting the four factors described by Woods (2003) as important challenges for managing a system's resilience. While it can be concluded that these factors negatively influence the ability of a socio-technical system to perform resiliently, it is not possible from this limited case study to claim the opposite, that is, to determine that resilient performance emerges simply by reversing these factors. Rather, the main contribution of this chapter was to present additional insights into the challenges facing the design of resilient systems. In particular, the results from this study demonstrated that double binds, negotiations, precedents, and not least, power relations between the various actors, play essential roles. In order to design resilient socio-technical systems, these aspects need further consideration, and the chapter underlines the importance of paying attention to the multi-actor context in which socio-technical systems are designed and managed.

## Concluding Remark

While the various activities and processes going on at the level of individual teams and organisations have gained considerable attention in the field of resilience engineering, less focus has been directed towards the effects these activities have on the potential for resilient performance of the higher levels of a socio-technical system. The case study presented in this chapter showed that the system's ability to perform resiliently at the regional and national level was constrained by the local perspective adopted by the various stakeholders. This chapter has thus contributed by emphasising the need to take the cross-scale interactions between various levels of a socio-technical system into consideration by illustrating the ways in which these interactions poses challenges to designing resilient systems.

## Commentary

This chapter shows how the basic arguments presented by Komatsubara (Chapter 7) can be applied in a more formalised manner to a concrete example. It has provided an illustration of the many dependencies and potential conflicts that exist in the real world, and that have to be taken into account when trying to manage large projects. The temporal relations, in particular, are often very important but difficult to see, especially if safety management relies on a structural description of the organisation. The overall functioning, and therefore also the resilience, should not be considered without acknowledging the intricate interplay between the multiple stakeholders that exist across levels and contexts.

## References

Birkland, T.A. and Waterman, S. (2009). The Politics and Policy Challenges of Disaster Resilience. In C.P. Nemeth, E. Hollnagel, and S. Dekker (eds), *Resilience Engineering Perspectives, Volume 2: Preparation and Restoration*. Farnham: Ashgate Publishing Limited, 15–38.

Cedergren, A. (2011). Challenges in Designing Resilient Socio-technical Systems: A Case Study of Railway Tunnel Projects. In E. Hollnagel, E. Rigaud, and D. Besnard (eds), *Proceedings of the fourth Resilience Engineering Symposium*. Sophia Antipolis, France: Presses des MINES, 58–64.

Cedergren, A. (2013). Designing resilient infrastructure systems: a case study of decision-making challenges in railway tunnel projects. *Journal of Risk Research*, 16(5), 563–582. doi:10.1080/13669877.2012.726241

de Bruijne, M., Boin, A. and van Eeten, M. (2010). Resilience: Exploring the Concept and Its Meanings. In A. Boin, L.K. Comfort and C.C. Demchak (eds), *Designing Resilience: Preparing for Extreme Events*. Pittsburgh: University of Pittsburgh Press, 13–32.

Dekker, S. (2006). *The Field Guide to Understanding Human Error*. Aldershot: Ashgate Publishing Limited.

Hale, A. and Heijer, T. (2006). Is Resilience Really Necessary? The Case of Railways. In E. Hollnagel, D.D. Woods and N. Leveson (eds), *Resilience Engineering: Concepts and Precepts*. Aldershot: Ashgate Publishing Limited, 125–147.

McDonald, N. (2006). Organizational Resilience and Industrial Risk. In E. Hollnagel, D.D. Woods and N. Leveson (eds), *Resilience Engineering: Concepts and Precepts* (pp. 155–180). Aldershot: Ashgate Publishing Limited, 155–180.

Mendonça, D. (2008). Measures of Resilient Performance. In E. Hollnagel, C.P. Nemeth and S. Dekker (eds), *Resilience Engineering Perspectives: Remaining Sensitive to the Possibility of Failure*. Aldershot: Ashgate Publishing Limited, Vol. 1, 29–47.

van Asselt, M.B.A. and Renn, O. (2011). Risk governance. *Journal of Risk Research*, 14(4), 431–449.

Vaughan, D. (1996). *The Challenger Launch Decision: Risky Technology, Culture, and Deviance at NASA*. Chicago: The University of Chicago Press.

Woods, D.D. (2003). Creating Foresight: How Resilience Engineering Can Transform NASA's Approach to Risky Decision Making. Testimony on The Future of NASA for Committee on Commerce, Science and Transportation. John McCain, Chair, October 29, 2003. Washington D.C.

Woods, D.D. (2006). Essential Characteristics of Resilience. In E. Hollnagel, D.D. Woods and N. Leveson (eds), *Resilience Engineering: Concepts and Precepts*. Aldershot: Ashgate Publishing Limited, 21–34.

Woods, D.D., Schenk, J. and Allen, T.T. (2009). An Initial Comparison of Selected Models of System Resilience. In C.P. Nemeth, E. Hollnagel and S. Dekker (eds), *Resilience Engineering Perspectives, Volume 2: Preparation and Restoration*. Farnham: Ashgate Publishing Limited, 73–94.

# Chapter 9
# Some Thoughts on How to Align the Theoretical Understanding of Team Performance with Resilience Engineering Theory

Johan Bergström, Eder Henriqson and Nicklas Dahlström

## Introduction

Recent contributions to the field of Resilience Engineering (RE) have added to the continuous development of new concepts and methodologies to improve resilience at different organisational levels. Part of these contributions has focused on training for adaptive capacity of individuals and teams to cope with changes and disturbances of work, since literature recognise that working tasks (at least in complex settings) are not as stable as procedures, manuals and regulations might depict. It is becoming accepted that more is needed than training for recognition of pre-defined situations and application of corresponding procedures, that is, individuals and teams should be prepared by their training to also cope with unexpected situations. In previous volumes of RE contributions we have introduced new methods in order to address these unexpected situations (Bergström, Dahlström and Petersen, 2011; Dekker, Dahlström, van Winsen and Nyce, 2008). In this volume we will rather discuss the theoretical foundation of team training and the potential to align such a foundation with RE theory. Guided by two of the four cornerstones of RE

(Hollnagel, 2011) our argument is that traditional approaches to sharp-end training should be reviewed, revised and readapted to concepts more aligned with RE thinking.

This chapter will initially discuss the traditional perspective on high-risk industry team training in order to present some of its discordance with resilience concepts. Then an alternative approach to such training is suggested. The suggested approach connects the RE abilities of responding and monitoring (Hollnagel, 2011) to the theoretical perspectives of distributed cognition and complexity. The aim of outlining a theoretical agenda is for future development of methods for training and assessment, of team performance in high-risk industries, to be more aligned with RE theory.

## Traditional Training Principles

An important starting point to understand the current paradigm of traditional high-risk industry team training is that it is based on the notion of task stability. Since technical systems are assumed to be highly predictable, and humans provide a potentially important but unpredictable and fallible variability to the system, proceduralisation of work is central. Consequently, training is focused on identification and reaction, that is, a work situation is identified and the corresponding procedures for the situation are executed. In this sense, the human operator is seen as an effective repository for a large number of stimuli-response programs for pre-defined situations (Hollnagel and Woods, 2005).

The study of man as a stimuli-response repository puts the emphasis on human cognition, which is supposed to produce correct and rational decisions according to optimal information processing, based on accurate awareness of the situation, assertive communication and effective interaction between leaders and followers. The ultimate implication of this information-processing paradigm, currently and traditionally dominant in the field of Human Factors and team training, is the use of different techniques to assess cognitive processes by means of behavioural analysis (Flin, O'Connor and Crichton, 2008). The central idea is that human behaviour is a genuine representation of human cognition, since certain cognitive processes should result in certain behaviours. This has led to an increasing use of evaluation techniques based on behavioural markers, such as NOTECHS and different forms of on-the-job observations focused on error identification and categorisation, such as Line Operations Safety Audit (LOSA).

But it is not only the focus on human cognition and behaviour that constitute the traditional high-risk industry training paradigm. In recent years, an increasing importance has been given to the use of scenarios based on previous incidents. In this case, past incidents are brought into training in order to 'fix' a perceived problem that caused those incidents to occur. In most cases this training reinforces existing procedures or introduces additional procedures intended to prevent the same incident from reoccurring. This again assumes stability; not only of tasks but also of risks, assuming that increased system rigidity can indeed eliminate discovered risks. International Air Transport Association (IATA), for example, have recently, as a part of their training and qualification initiative, brought together several of the above-mentioned concepts under the umbrella of 'Evidence Based Training' (EBT). EBT emphasises a need for gathering, sharing and categorising data from past events in order to have the lessons from them incorporated in future training (Voss, 2012). Some of the first conclusions drawn from the EBT project highlight issues related to pilots' attitude toward safety, such as problems related to non-compliance with standard operating procedures even in threatening circumstances. Also the accident with Air France 447 has triggered central stakeholders to draw similar conclusions, including The European Safety Agency (2012) and the UK CAA (2013) publishing accounts which reproduce what is considered 'acceptable' explanations, depriving the industry from far more difficult and complex explanations. Such explanations would allow an increased understanding of the role of the operator in the system and direct how this role could be strengthened by a focus on resilient behaviour rather than on the stimuli-response behaviour of procedural adherence.

Although many of these models and techniques are simple to understand and convenient in how they ascribe safety-related problems in the operational environment to the behaviours of the individuals operating in such an environment, our argument is that this simplicity represents a considerable trade-off, that is, there is a price to pay. While these models and techniques normatively assume task stability, they effectively exclude and hide the complexities of numerous highly contextual factors in the work situation. If these factors would instead be identified they could contribute to operator understanding of the inherent risks in their work processes and enhance their resilience in handling the variability in day-to-day situations as well as their handling of unusual and unexpected situations.

## The Resilience Approach

Hollnagel (2011) explains that the approach of RE is focused on an in-depth understanding of an organisation's ability to 'adjust its functioning prior to, during or following changes and disturbances, so that it can sustain required operations under both expected and unexpected conditions' (p. xxxvi). This broader definition is then operationalised as the study of 'the abilities to respond to the actual, to monitor the critical, to anticipate the potential and to learn from the factual' (p. xxxvii). The first question in order to develop a framework for a resilience approach to sharp-end training becomes which of these four abilities that can at all be addressed in training. In this chapter we will focus on outlining a conceptual agenda for how team training can be used to enhance the abilities to respond to the actual and monitor the critical, with the abilities to anticipate and learn seen as potential secondary effects of such training. Underlying this conceptual agenda will be the perspectives provided by cognition as a phenomenon distributed among the actors engaged in a specific context (Hutchins, 1995a) as well as complexity theory. Together these two perspectives embrace the context-specific nature of monitoring and response as well as how both these abilities are emergent properties of multiple relations and interactions rather than the products of reliable behaviour at the level of the individual actor.

Complexity theory embraces the corollary that a system cannot be fully described nor fully controlled (Cilliers, 2005), which is fundamentally important for understanding safety-critical work from an RE perspective. Complexity is what emerges when a system, put together by physically separated actors, elements and artefacts (for example, different wards at a hospital or different aircraft on approach to an airport), shifts from loose to tight coupling and from high autonomy to high interdependence in a short span of time, for example, as when different hospital wards normally functioning relatively autonomously become highly interdependent in response to an escalating situation (Bergström, 2012; Dekker, 2005). Likewise, instead of training only for task stability and 'correct behaviour', the RE approach needs to embrace that variability is not only normal, but also necessary for the ability to dynamically adapt to unexpected situations. This does not mean that all procedures are detrimental to safe work, but that understanding of procedures and their application in a variety of predicted situations, as well as skills for handling of unexpected situations, must be the focus of training.

From the approach of analysing cognition not as individual-centred but rather as a distributed trait of a situation, the focus shifts from the human as an information processor to the actual work in which the human engages, together with technological systems and other humans (Woods, 2003). In this sense, the unit of analysis is changed from the mind of an individual to the joint system of humans and artefacts engaged in a particular working situation (Hutchins, 1995b).

Thus, from the perspective of RE the understanding of the ability to respond and monitor lies neither in observing human behaviour, nor in deconstructing human work from engineering-centred task analysis models, motivationally based models or concepts such as workload management, situation awareness and decision-making. Instead the interest of analysis lies in the complexities facing the sharp-end operators in their day-to-day work and how to improve their adaptive capacity in order to promote safety.

## Important Principles for Training Response and Monitoring

In our respective organisations we are incorporating the notion of resilience into training programs as well as in methods to assess team performance. Bergström is involved in developing a training program for multi-professional team training of healthcare staff.

Henriqson and Bergström are together developing methods for team assessment based on the theoretical fundaments outlined here, and Dahlström has, together with Bergström, worked for several years with alternative approaches for training team coordination in escalating situations. Dahlström is in recent years closer than ever to 'reality', being the Human Factors Manager at Emirates Airlines Flight Operations Training. With the context of our work outlined, the last part of the paper will be focused on describing the main principles of the resilience approach to team training and team performance assessment.

*Responding to the Actual*

Just as concepts such as success, safety or risk, the performance of teams needs to be understood as an emergent property of a multitude of interactions and relations, embracing that 'there are no fundamental differences between performance that leads to failures and performance that leads to successes' (Hollnagel, 2011, p. xxxv). This should be seen as a contrasting view to the idea that team performance can be ensured by best-practice guidelines or by gathering increasingly more data on their past performance. This tenet also questions the idea that concepts such as 'correct behaviour' can be imposed on the system. To move beyond containing or limiting the emergence of variance in performance, by means of stronger emphasis on following standard operating procedures or replacing unreliable humans with reliable technology, the resilience approach asks us to establish practices for dealing with variability and uncertainty. In this sense, in order to analyse the capacity of individuals and teams to perform work, or to analyse the design of technology, it is necessary to focus on the phenomena emerging from the interactions of joint cognitive work (rather than on delimited individual cognition and action).

The interactive and dynamic context of normal work can offer guidance for action at the same time as it will be affected by the action carried out. In this case, there may be no precedence in regards to specific modes of interaction, except for those required for immediate needs of the event (for example, for in-flight collision avoidance). For example, Henriqson, Saurin and

Bergström (2010) identified that local representations, which condition the broader coordination context in a cockpit, are results from interactions between internal representations (interpretative structures unique to the individual) and external representations (elements in the action context, such as symbols, numbers, data and shapes), which are in themselves always partial and incomplete. This means that rather than only providing finalised action programs (that is, procedures) for events, it is imperative to ensure operator understanding of the multitude of available representations and interactions and develop the skills to apply these to a variety of situations.

Just as Nyssen (2011), we emphasise that the joint cognitive systems-notion of coordination forms an important part of the framework in understanding organisational response. In the study mentioned above (Henriqson, Saurin and Bergström, 2010) we describe how coordination can be interpreted as a situated and distributed cognitive phenomenon in the cockpit of commercial aircraft. The study provides an integration of the perspectives of joint cognitive systems theory with four coordination requirements described in the literature: common ground, interpretability, directability and synchrony (Klien et al., 2005). We have also used these coordination requirements in studies aiming at interpreting joint cognitive work in escalating situations, forming an early suggestion for a method to assess team performance based on joint cognitive systems theory (Bergström, Dahlström, Henriqson and Dekker, 2010).

When developing programs for team training the importance of the multi-professional approach is stressed. In a multi-professional setting the participants can together identify the situations in which the relationships and interactions between them, rather than their respective reliability as safe components, establish safety (or risk) as an emergent property of the system. This multi-professional dialogue is seen not only as a vital activity for enhancing the ability of the organisation to enhance future response to changing conditions and monitor its current performance, but also for widening the learning-loops in the organisation and anticipate the characteristics of future interactions.

Furthermore, joint cognitive systems theory has adopted a cybernetics approach to define control by its circularities of feedback and feed-forward. This approach combines the cybernetic notion of regulation (Ashby, 1959), the Perceptual Cycle of Neisser (1976) and Hutchins' ideas of distributed cognition (1995a, 1995b) to provide a functionalist approach of control. In this sense, control 'happens' during the interaction between 'human-task-artefact' and is goal-oriented and influenced by the context in which the situated activity takes place. From the resilience perspective outlined here we can also see control as an emergent property of the systems' ability to respond. We have been involved in developing methods for team performance assessment through operationalisation of Hollnagel's Contextual Control Model (Hollnagel and Woods, 2005) in order to map how the interpreted level of control shifts during a scenario by including both the participants' own reflections and the observer's interpretations (Palmqvist, Bergström and Henriqson, 2011). Noteworthy is that a contextual model of control is, in contrast to behavioural markers or error categorisations, not normative, meaning that no level of control is seen as more appropriate than another but instead as highly contingent upon the situation and context. This is a promising way to go beyond the behavioural markers or error classifications of the information-processing paradigm.

### *Monitoring the Critical*

Even a widely accepted information processing notion such as 'situational awareness' becomes problematic from the resilience perspective. Complex systems do not allow for complete or any other finalised and 'correct' descriptions. Instead of using hindsight to accuse operators for not having had optimal situation awareness and trying to improve this by training of 'correct monitoring', the focus of training needs to shift to ensuring that different and competing productions of meaning (based on different experiences and viewpoints) are available when working in safety critical environments. This can allow a parallel development of best practice for routine situations as well as skills development for unusual and unexpected situations.

Ultimately, this is a shift towards the notion of diversity, which is fundamental for complexity theory (Cilliers, 1998). Complex systems are resilient when they are diverse, which also needs to be embraced by the resilience interpretation of the ability to monitor. Diversity implies that different practitioners deploy different repertoires for responding to what, from their respective perspective, is seen as evidence as well as to each others' constructions of such evidence (Dekker, 2011). This argument was also made in a recent study analysing the joint interactions of the multiple professionals (especially midwives) involved in the process of labour. The study raised questions regarding the idea of understanding performance by reference to best practice guidelines, emphasising that:

> Patient safety efforts, then, might recognise, celebrate, and enhance the positive aspects of diversity that guarantee the emergence of resilience in complex situations. Such efforts can be made in activities of inter-professional team training where medical staff representatives are given the opportunity to identify complex as well as complicated situations in their work to achieve more efficient and effective patient-centred care (Dekker, Bergström, Amer-Wåhlin and Cilliers, 2012).

From the resilience perspective the notion of monitoring needs to be raised to a level of a joint cognitive effort, facilitated by a focus on strategies to monitor and bridge the current gaps in the system (Cook, Render and Woods, 2000). Again, when developing concepts for multi-professional team training in complex settings it is important to gather the actors that at certain points might become tightly coupled and highly interdependent (Bergström, Dekker, Nyce and Amer-Wåhlin, 2012). Together they need to be given the opportunity to build common ground, aided by facilitators, and enrich their different perspective with those of others in order to create prerequisites for successful response at the organisation's most complex moments.

## Summing Up the Argument

Resilience Engineering theory embraces how successes and failures are the results of the same kind of processes in dynamic and goal-conflicted environments. This important principle challenges traditional practices that assume a task-stable environment. Consequently, in efforts to enhance organisational resilience, there is a need for the establishment of a theoretical agenda to team performance and training which is more aligned with RE theory. This chapter discusses some thoughts for what to include in such an agenda.

In order to be aligned with the central notions of RE theory we argue that concepts to enhance the organisational abilities of responding to the actual and monitoring the critical need to be rooted in complexity theory and the theory of distributed cognition. The two perspectives complement each other. The perspective of distributed cognition helps to shift the unit of analysis from the individual actor to the coordination activities of the joint cognitive system. These activities can be understood and analysed based on certain requirements for successful coordination and/or the level of control emerging from the coordination activities. With the concept of emergence being essential to complexity theory, here is also where this perspective becomes important. Not only does complexity theory suggest that the unit of analysis needs to be the interactions and relations between actors rather than the behaviour of the individual actor, but it also helps outline the agenda for how to monitor the critical by the tenet that the complex system is resilient when it is diverse. Consequently the agenda for team training needs to be one of emphasising and enhancing diversity rather than, as is the risk of traditional approaches to team training, uniformity and rigidity.

The multi-professional discussion is seen as important to enhance organisational diversity and here the third and fourth cornerstones of RE as outlined by Hollnagel (2011) come in: the notion of anticipating and the notion of learning. It is in this multi-professional discussion that actors are able to learn about each other's perspectives in order for them to establish the common ground which might make them able to anticipate the current and future actions of the joint cognitive system in which the actors engage. From this perspective the four activities are not disconnected or isolated when it comes to team training, but rather all four are tightly intertwined activities and can benefit from such multi-professional learning.

## Commentary

Safety management, whether resilient or not, obviously depends on the competence and experience of the people working in the system. This competence and experience is to a large extent based on the training that is provided. This chapter shows how the fundamental premise of Resilience Engineering (that tasks are not stable and that variability and adjustments therefore are necessary) has consequences for training. The chapter presents a discussion of the requirements for a theoretical agenda for team performance and training that is aligned with the principles of Resilience Engineering, especially the abilities to respond and to monitor. Developing such an agenda is an important step from requirements to specifications.

# Chapter 10
# Noticing Brittleness, Designing for Resilience

Elizabeth Lay and Matthieu Branlat

Engineering is the discipline of applying art or science to practical problems. Resilience Engineering is the discipline of applying principles of Highly Reliable Organizing and Resilience Engineering to the design of resilient systems. An assessment of brittleness is the first step in determining which strategies and tactics to deploy and noticing brittleness (and resilience) is a skill that can be learned. This chapter covers how the skill to notice brittleness can be developed then applied in a workshop to assess brittleness and subsequently design strategies and tactics to increase resilience. These topics will be explored in the context of maintenance work.

## Introduction

Organizations operating in high-risk/high-consequence domains recognize the variability of their work environment and its potential consequences on performance and safety. As a result, they actively seek ways to deal with this variability in order to avoid undesired states and outcomes. Traditional approaches based on risk management aim at anticipating, measuring and building mechanisms to address specific forms of variability often with a goal of reducing variability.

Although such approaches have shown positive results through building basic adaptive capacity within the systems considered, they are also based on strong assumptions that make the systems ineffective at managing disruptions; they typically overestimate their knowledge of the various forms of variability (oversimplified models of the world), and collapse in the face of surprising events. Resilience Engineering (RE) represents a different type of approach: rather than anticipating specific events, it assumes that the world is variable, that this variability cannot always be known in advance, and that it might even be surprising. As a result, RE aims at describing and designing the mechanisms that will support systems' adaptive capacity in the face of known and unknown variations in the world: mechanisms that allow for the detection of the variability, for the understanding of its potentially surprising nature or scope, and for the timely reconfiguration of the system to manage it successfully. While characteristics of already existing High Reliability Organizations (HROs) and resilient systems have been thoroughly described, purposefully engineering resilience into a system is uncommon. Questions arise about what practical transformations can be made in support of resilience, and also about how to conduct interventions that introduce RE principles in organizations that have a more traditional risk management culture.

This chapter aims at describing our experience with these very issues, and at illustrating them through specific interventions in the domain of industrial maintenance.

## Underlying Principles

### *Resilience and Brittleness*

Resilience is the intrinsic ability of an organization to adjust its functioning prior to, during, or following changes and disturbances, so that it can sustain required operations under both expected and unexpected conditions (Hollnagel, 2012). Resilient systems are agile in the face of change and have buffers (margins of maneuver) to respond to unforeseen demands, thus creating the conditions to avoid or minimize consequences of adverse events; they understand and manage their complexity. Brittle systems, on the other hand, fail to notice warnings and to adjust their behavior in time to prevent collapse; they overlook and fall victim of tight couplings. They may have a system designed around "standard" (low variability) maintenance even though variable scope is the norm ("standard" never happens).

Brittleness and resilience are two sides of the same coin. Just as where there are threats, there are corresponding opportunities; where an organization or system (or process) is brittle, there exists the possibility to increase resilience. Brittleness and resilience are system properties not outcomes (Cook, 2012). You can have a good outcome with a brittle system or a bad outcome with a resilient system although the probability of having a desired outcome increases with resilient systems. "Resilience Engineering ... agenda is to control or manage a system's adaptive capacities based on empirical evidence;" "to achieve resilient control ... system must have capacity to reflect on how well it is adapted, what it is adapted to, and what is changing in its environment." Managers need knowledge of how the system is resilient and brittle to make decisions on how to invest resources to increase resilience (Woods, 2006; Hollnagel, 2009). "Resilience/brittleness of a system captures how well it can adapt to handle events that challenge the boundary conditions for its operation" (Woods and Branlat, 2011).

*Resilience as Management of Variability and Complexity*

Work systems, as Complex Adaptive Systems, require adaptability in order to be resilient when anomalous situations arise, that is, maintain sufficient levels of safety and performance in the face of disruptions. However, adaptive processes can be fallible: systems may fail to adapt in situations requiring new ways of functioning; or the adaptations themselves may produce undesired consequences, especially as a result of unmanaged functional interdependencies. These challenges have been abstracted through three basic patterns in how adaptive systems fail (Woods and Branlat, 2011).

The three basic patterns are

1. Decompensation – when the system exhausts its capacity to adapt as disturbances/challenges cascade. This pattern corresponds to situations where the system is unable to transition to new modes of functioning in a timely manner

to respond to the disturbances.

2. Working at cross-purposes – when roles exhibit behaviour that is locally adaptive but globally maladaptive. This pattern is a result of mis-coordination across the system and corresponds to a failure to manage functional interdependencies; issues can arise from interdependencies that remain undetected until they are revealed by incidents following a disturbance.
3. Getting stuck in outdated behaviours – when the system over-relies on past successes. This pattern results from organizations failing to revise their models and plans in place, often through oversimplifying or disregarding the disturbances they experience.

These patterns of failure propose a description of how an organization is unsuccessful at managing adverse events, and suggest ways to transform the system so that it better manages its complexity in a variable environment. Measures to increase resilience derive directly from the nature of the patterns that they counterbalance. Corresponding forms of improvements include: (1) management of resources, for example, by creating tactical reserves and understanding the conditions for their employment; (2) coordination within the system, for example, by investigating and supporting functional interdependencies; (3) learning mechanisms, for example, by transforming the models underlying the investigation of incidents.

### *Resilience Compared with Traditional Risk Assessment*

Traditional risk assessments typically include identifying risks (specific and detailed), analyzing the risks (qualify, quantify, rank), and designing specific responses for higher ranked risks. Risk assessment often focuses on preventing things that can go wrong.

Resilience Engineering involves designing to ensure things go right with more focus on preparedness and less on prediction. Consideration is given to broad, big picture situations and uncertainties. Responses tend to be general versus specific. Resilience Engineering involves:

- Bounding uncertainty and possible outcomes: considering properties of systems when analyzing possible failures; understanding fundamental limitations of resources; describing possible big picture outcomes; and seeking to be approximately right across a broad set of eventualities (Taleb, 2010).
- General response design: making the distinction between positive and negative contingencies; seizing opportunities; investing in preparedness, not prediction (Taleb, 2010); since variability is inevitable, looking for ways to build structure around and plan for variability; and designing general responses that could address a broad set of situations (without focusing on the precise and local).

Uncertainty is defined as the state of having limited knowledge where it is not possible to exactly describe existing state or future outcome, it is an unmeasurable risk and includes what we don't know, ambiguity, and/or variability. In project management, according to De Meyer, there are four types of uncertainty (De Meyer et al., 2002, 61–2):

- Variation: a range of values on a particular activity.
- Foreseen uncertainty: identifiable and understood influences that the team cannot be sure will occur.
- Unforeseen uncertainty: can't be identified in planning, team is unaware of event's possibility or considers it unlikely. Also called "unknown unknowns."
- Chaos: Even the basic structure of the plan is uncertain. There is constant change, iteration, evolution. Final results may be completely different from original intent.

During risk assessment, the tendency can be to act like the future can be more accurately predicted than is possible, such as when probabilities are estimated to a high degree of granularity. In risk assessment, uncertainty may be neglected. This is due, in part, to the psychological make-up of humans; studies have shown people are more averse to uncertainty than to risk alone (Platt and Huettel, 2008, 398–403). To be highly resilient is to be prepared for uncertainty. To be highly resilient is to respond robustly to the unexpected.

## Observing Brittleness at Play

Noticing brittleness and resilience is a skill that can be learned. One approach to building this skill is through study groups by: understanding the principles through literature, observing and recognizing the patterns and characteristics of resilience and brittleness in everyday work, then discussing observations across situations and domains.

Over a period of time, people build the skill to notice brittleness and resilience by employing reciprocation (conversations), recurrence (periodic conversations and observations), and recursion (making repeated observations as knowledge is built to deepen understanding).

According to Weick and Sutcliffe, organizations are brittle if they (Weick et al., 2001):

- Have little or no reserve
- Don't pay attention to or deny small failures
- Make assumptions, small misjudgments
- Accept simple diagnosis, don't question
- Take frontline operations for granted
- Defer to authorities rather than experts
- Keep working as usual upon disruptions.

The following table describes additional and expanded signs of brittleness that have been observed in the context of complex, variable industrial maintenance work situations. Relationships with the 3 maladaptive patterns (1–decompensation; 2–working at cross-purposes; 3–getting stuck in outdated behaviour) are indicated between parentheses.

## Table 10.1 Observations of brittleness at play

| Type of sign | Examples of observations |
|---|---|
| Buffers/reserves | No buffer or contingency plans for critical sequential events (1)<br>Critical singular resources (only one person with skill, only one tool), more brittle if long lead time to procure (1)<br>Single point failures, in general (being reliant on one vendor for critical work or resources) and especially serial activities with potential single point failures (1, 3)<br>Over use – burn-out – of key personnel. (2) |
| Stiffness/rigidity/lacking flexibility | Fixed configuration teams made of highly specialized, singularly skilled workers (especially if there is little reserve and multiple skills are necessary for work) (1, 2, 3) |
| Information and knowledge | Not knowing which resources are critical. Having low reserves of critical resources with a variable demand for those resources (people, tools, materials, supplies, etc.). A "critical resource" is defined as a resource on which critical path work is dependent.<br>Leadership lacks big picture view (2)<br>Leadership not having ability to get current information that describes changing situation (1, 2)<br>Lack knowledge of how changes (both large & many small) impact big picture or program (2, 3)<br>Communication is not frequent and timely beyond yield point (begin to lose control) (1, 2,) |
| Variability and uncertainty | Not exploring or planning for where uncertainty lies (3)<br>Not understanding dependencies and interactions (competition for same resources) (2, 3)<br>Ungrounded or unlikely assumptions (3)<br>Lack bounding variability (most likely and worst case scenarios) (3)<br>1st or 2nd time use of critical process, supplier or uncertainty that is added from not having experience or history, in general (3) |

**Table 10.1** *Concluded*

| Type of sign | Examples of observations |
|---|---|
| Planning | Lack plan for monitoring including where yield (begin to lose control) and failure points lie (1) |
| | Not exploring or planning for potentially disruptive risks (technical issues) (1, 3) |
| | Late assignments of resources (no time for them to plan/prepare) (1) |
| | Lack of coordinated planning for big picture or group of projects that require inputs from several organizations or teams (lack shared resource planning model) (1, 2) |
| | Highly likely, potentially disruptive emergent scope not included in plan even though history shows this to be common. (3) |
| | Lack consideration of timing of disruption/issues on big picture impact (Which issues likely to arise early and significantly disrupt downstream projects? Where at risk for early disruptions?) (2, 3) |
| | Step change in demands on resources wherein there is no change in the way planning or managing occurs, operations continue as normal (3) |
| | Over reliance on "fire-fighting". Last minute, high levels of change with little time to react (Plan "breaks down" after first tranche, teams start together but are soon fragmented. Effects of fragmentation not taken into account in planning or design of teams. Planning not performed to minimize fragmentation.) (1, 2) |

## Implementing Principles Of Resilience Engineering

This section describes the structure and content of a workshop that could be held for the purpose of introducing principles of Resilience Engineering in an organization and identifying ways to increase operations' resilience. This description, which aims at providing a general guideline, is based on the conduction of such events in the context of high-risk/high-consequence industries. One of the central themes of the workshops conducted (and of the example described in this section) is the management of situations where load or demand potentially exceeds capacity of existing resources. Such focus is operationally relevant across industries and resonates with key questions about how to design for resilient control (Woods and Branlat, 2010). The design is based on the following properties for increased resilience (Woods, 2006, 23):

- buffering capacity: system ability to absorb disruptions without breaking down;
- flexibility versus stiffness: ability to restructure in response to changes;
- margin: how close system is operating relative to performance boundaries (operation with little or no margin is precarious)
- tolerance: system behavior near boundaries; degrades gracefully or collapses.

*Workshop Participants*

Ultimately, the goal of the workshop is to leverage participants' work experience to help them diagnose their organizations' brittleness and resilience, as well as to identify potentially fruitful directions to increase resilience. In this collaborative problem-solving type of task, the general approach is to generate success through diversity rather than through a few high performers (Hong and Page, 2004):

- cross-domain representation brings diversity in perspectives and a fuller spectrum of relevant concerns;
- consider including operations, engineering, marketing, project managers, resource planners (parts, people, and tools), and other groups that come in contact with or influence the project.

One of the most important components of the workshop is the facilitator. To help with the design of resilience, the conversation has to be led by a person who has the skills to notice brittleness and resilience, ideally supported by operational experience. Even though the workshop is structured, holding such a workshop is more complex than following a series of steps; it is critical that the facilitator has the abilities to notice and probe risk, uncertainty, and brittleness.

*Helping Participants Notice Brittleness*

A suggested agenda is to begin with introducing Resilience Engineering, following with an overview of the situation, current plan, and identification of key issues and concerns. As issues and concerns are raised, it is the role of the facilitator to begin to identify and note areas of brittleness for the group. This builds the foundation from which to perform a deeper diagnosis of brittleness using questions designed around HRO principles and properties of resilient systems. See Table 10.2 for sample questions. Assumptions in planning are rich fodder for finding brittleness; its likely surprises will be found within them.

Diagnosing brittleness involves questioning into and probing:

- boundary areas and where performance began to degrade (margins/yield points/failure points)
- specific projects for possible effects on programs
- areas of variance, ambiguity, and assumptions
- boundaries, limitations, critical resources, reserves, and other constraints
- key decisions that can be changed or have not been yet made
- interactions among risks for possible cascading situations.

Once significant risks, uncertainties, and critical scenarios are identified, these should be elaborated in more detail and bounded such that they can be responded to. The details of which project and exactly what the work is may not be important. It may be enough to know that if some type of emergent work occurs with this general timing, it could have this general impact. That can be enough to enable design of a flexible, general response to reduce disturbances such as moving people in the midst of a project.

**Table 10.2 Sample workshop questions**

| |
|---|
| Where do you lack information, which projects, processes, plans are not well defined (sources of uncertainty)? |
| What critical decisions have yet to be made? |
| What assumptions have you made? |
| What is new, novel, or different that adds risk or uncertainty? |
| Has anything changed that makes these issues more likely to cause failure? |
| Where is there uncertainty due to operation or maintenance history? |
| Is there anything you are uncomfortable with? |
| What constrains you in your ability to execute? |
| What will "stretch" or "stress" our system? Who will be most heavily loaded/stressed? |
| What combination of small failures could lead to a large problem? |
| Where can we easily add extra capacity to remove stressors? |
| What can we put in place to relieve, lighten, moderate, reduce and decrease stress or load? |
| Will there be times, such as during peak load, when we need to manage or support differently? What is the trigger? |
| Which support organizations need to be especially sensitive to front line needs and what is our plan to accomplish this? |

Pinging is defined as the proactive probing for risk profile changes (Lay, 2011). A "pinging plan" identifies the triggers that prompt moving into different actions and supports monitoring. The plan includes identification of yield (things begin to fall apart) and failure points (run out of what, when) and margin (space between where you are and failure point).

**Table 10.3 "Straw man" for pinging design. Indicators that risk level is increasing**

| Risk Level Indicator | Green | Yellow | Red |
|---|---|---|---|
| Scope expands by x | X1 | X2 | X3 |
| Inspection finds | Description 1 | Description 2 | Description 3 |
| Schedule extends | 0 days | 1–2 days | > 2 days |
| Customer relationship | Working as team, good communication | Tense, some communication breakdowns | Conflictive, lack trust, poor communication |
| # of significant issues team is dealing with simultaneously | <2 | 2–3 | >3 |
| Human resources | Fully staffed, majority rested, no change out of people | Short 1–2 people, some fatigued, change out 1–2 workers during project | Short > 2 people, many fatigued, change leads out, change out >2 workers, critical functions missing or late |

## Designing to Increase Resilience

According to Hollnagel (2012), resilient organizations: learn from history; respond (adapt) to regular and irregular conditions in a flexible, effective manner; monitor short-term threats and opportunities; revise risk models; and anticipate long-term threats and opportunities. Strategies and tactics related to the capacity of human resources being at or near the limit are shared below. They are grouped in tables according to themes. Rather than recipes, these tables provide guidelines that need to be tailored to the specificities of the situations considered. Details are provided to illustrate how specific strategies could be designed. They are based on experience in the domain of industrial maintenance.

## Respond (Adapt)

| Type of strategy | Examples |
|---|---|
| Management of deployed resources | Shift goals, shift roles, have critical resources perform critical tasks, only! Use less experienced people for less complex work; provide more oversight if needed ... experts coach, provide oversight versus "do" work.<br>Add buffer such as a logistics person to manage parts, people, tools; especially for projects with multiple emergent work scopes or issues; commercial or other support to free project manager to focus on managing the job, or a human performance/ risk/safety/quality specialist to perform additional checks and bring outside perspective. |
| Provision of extra resources | "Drop in an Expert". This concept involves finding a person with deep, relevant knowledge (possibly a retiree) and funding them for a short time period with the mission to assess the situation then make offers to groups who need help. Both the expert and the groups who need help determine where the expert can offer the most help.<br>Form a crisis management team, typically made up of managers, to bring about a heightened state of coordination and help. Consider the decisions that need to be made and the power needed to remove barriers and expedite solutions in determining team members. The team strengthens leadership's connection to the front lines and provides a forum for project managers to escalate issues to management's attention. The team can be more effective with authority to add or move resources as needed.<br>Form a dedicated rapid response team, typically made up of professionals. As risks and issues multiply, this team can be assigned full time to removing barriers and implementing solutions. A cross-organization group can improve collaboration and hold a neutral position to smooth political tensions that arise during periods of high stress. The focus should be to aggressively address issues that have the potential to delay front-line work.<br>Increase use of human performance tools. Consider which tools could be deployed that currently are not being used and how tools could be used more effectively, such as a defined plan for peer checks. |
| Management of priorities | Adjust capacity limits by removing stressors from people. Shed tasks: do only what's necessary, stop unnecessary work/paperwork.<br>Shed load: move, decline projects.<br>Manage differently considering how people respond when they are close to their limits (fatigued, stressed); they are more forgetful, less attentive and may miss things. |

The strategies described in the table above relate particularly to the first pattern of adaptive response described in a previous section. They aim at favouring timely response to events. Cook and Nemeth (2006) have described the successful management of mass casualty events by Israeli hospitals through similar strategies.

## Monitor

| Type of strategy | Examples |
| --- | --- |
| Support of processes of sense-making | Someone steps back from (or out of) their usual role to gain a broader perspective.<br>Begin a heightened state of coordination and help; possibly daily calls with those who are involved.<br>Assure communications occur with enough context to allow cross-checking.<br>Avoid the tendency to handle serially versus holistically. |
| Support reflective processes | Know where yield points are (what % deployed is sustainable?).<br>Look for signs the mood or situation has changed.<br>Stop and assess global situation: The water is coming up. Where is dike going to breach? Need to put reinforcements at breach points. Need global assessment of where things are going to come apart.<br>Identify where the breakdowns are occurring and brainstorm on where further breakdowns are likely to occur.<br>Query front lines on breakdowns, concerns, and current capacity.<br>Ask: Who is at the point they can't keep up? What resources or help is needed to add capacity, remove stressors, or free up capacity? What has affected or is impeding their ability to perform? What can be done to improve this situation? What can be done to unload workers or improve conditions? What is keeping them awake at night?<br>Continue to search for signs of brittleness, such as incomplete, unclear information or statuses, silo situations where workers were not optimally connected with front lines, communication issues, accuracy of assumptions, and key individuals for whom there is no back-up. |

Strategies described in the table above relate to how organizations assess and understand their situation. They address especially the second and third pattern described previously through the improving coordination (a source of information sharing) across the system, and through mechanisms aiming at bridging the gap between a situation as imagined and the actual situation experienced.

## Anticipate

| Type of strategy | Examples |
|---|---|
| Anticipate knowledge gaps and needs | Practice and build depth, before it's needed.<br>Develop multi-skilled workers. For example, a back office team that is also trained to hold various roles to unload or support front lines. A strategy for building this team is to recruit people with a variety of backgrounds with the understanding that they will periodically work the front lines to keep their skills fresh. Off peak, they could hold various support roles. |
| Anticipate resource gaps and needs | Anticipate losing people and their associated capacity.<br>Build buffering capacity and develop reserves before needed.<br>Design reconfigurable teams. This can be implemented by having a larger team that can be split into smaller components depending on the need, such as entire team working one shift or splitting to cover two shifts.<br>Pre-assign tactical reserves to planned work to reduce disturbances caused by emergent work. Tactical reserves could be back office personnel with appropriate experience. Assign them to planned work during peak load (giving them time to prepare), leaving active personnel available to respond to unplanned work with their more current skills enabling them to better handle variable situations. |

Strategies described above aim at supporting the processes of monitoring and response. They correspond to longer term learning processes and are responses to conditions experienced in the past that have hindered resilient operations.

## Conclusion

Probing interactions and interdependencies along with stepping back to look at the entire system differentiated this workshop from a traditional risk assessment. According to Erik Hollnagel, "... resilience and brittleness probably do not reside in components ... but are rather a product of how well the components work together. ... this must be understood by trying to comprehend the dynamics of how the system works, from a top-down rather than a bottom-up (component) view." He suggests that one should look to improve the everyday practices or the ways of working, instead of focusing on components of a system.

The principles of RE and HRO are not complex and are learned fairly quickly in a workshop setting. These principles suggest strategies which serve as a foundation for design of practices. The more difficult part is holding a shifted perspective while viewing issues that may not be new; a facilitator with knowledge in RE and HRO domains is crucial to accomplishing this. Per Erik Hollnagel (2012): "With resilience, noticing is different," and "A system's willingness to become aware of problems, is associated with its ability to act on them" (Ron Westrum, 1993). This might be the core of the real value of RE, gaining and holding a shifted perspective that enables a person to notice what we couldn't see before and, once noticed, we are compelled to act.

## Commentary

Classical approaches to safety management accept that the world is probabilistic, but also assume that it is stable in the sense that we can trust probability calculations. Resilience Engineering takes a different view, assuming that the world is variable but that the variability is not always known in advance. The variability is, however, orderly rather than stochastic. It is, therefore, possible to develop ways which will support a system's adaptive capacity vis-à-vis the variations in the world. By recognizing different types of uncertainty, the chapter illustrates how to go from theory to practice by showing how it was possible to teach people to be flexible and be prepared, and thereby avoid brittle performance.

# Chapter 11
# Sensor-driven Discovery of Resilient Performance: The Case of Debris Removal at Ground Zero, NYC, 2001

David Mendonça

*Abstract.* As has been amply demonstrated in numerous studies of post-disaster response and recovery operations, the instrumentation necessary for capturing data on resilient (or brittle) performance is rarely present when and where it needs to be. Yet because the organizations tasked with these operations seek to control them, they typically distribute a wide variety of "sensors" in the environment. An organization's decisions about how to collect and analyze sensor data are here viewed as providing a window into resilient performance and, more specifically, that decision-making processes that underlie it.

This chapter draws upon the case study of an ad hoc, largely autonomous organization formed to undertake debris removal at Ground Zero (New York City, NY) following the 2001 World Trade Center attacks. A central theme of this work is the attempt to triangulate incomplete or ambiguous readings from multiple sources, and to comment on what gaps in these readings tell us about the role of human cognition in achieving resilience. The case study is particularly relevant due to the acknowledged improvisatory nature of the response: indeed, the removal of debris from Ground Zero was a project unlike any other in recorded history, yet proceeded without any serious incident.

The chapter concludes with implications for post-incident studies of resilience in environments populated with ad hoc sensors.

## Introduction

Studies based on post-disaster field data offer opportunities for examining resilient performances in situations characterized by complexity, uncertainty, risk and urgency (Vidaillet, 2001). However, as has been well documented, particularly in the literature on human response to disaster (Dombrowsky, 2002), appropriate measurement instrumentation is rarely in place at the time of the disaster. Rarer still are the situations where pre-event data may be compared to post-event data. Opportunities for examining resilient performance in relation to hazards have traditionally been limited. First, the costs associated with large-scale and nearly continuous observation of pre-event conditions in human systems have been high. Second, the consequence of disasters can include destruction of established data collection instruments, as occurred with the emergency operations center and, later, at the New York Fire Department command post as a consequence of the World Trade Center attack. Third, new processes, technologies and personnel brought in to support the response may not be measurable with any instruments that remain available, as commonly occurs when the victims of a disaster act as first responders, or when ad hoc communication networks are formed. A very real challenge in the engineering of resilient performance is therefore fundamentally methodological: how can organizational theorists and designers develop and implement measurement instruments for "experiments" which are essentially undesignable?

Contemporary research on hazards must be cast in terms of radical changes in the data landscape over the past decade, including data proliferation, but also challenges of verification and validation (Dietze et al., forthcoming). A parallel line of sociological research (Murthy, 2008) has argued for deeper engagement with the vast body of data (so-called "big data" (Savage and Burrows, 2009)) being produced by both social and physical sensors (Buchanan and Bryman, 2007; Weick, 1985). In fact, this perspective harkens back to early mass observation studies (Simmel, 1903; Summerfield, 1985; Willcock, 1943), yet with human sensors largely replaced by technological ones. These technological artifacts are of course not theory-neutral: collection and synthesis of data from these sources—as with any other—must necessarily recognize the limits of measurement with which they are programmed (Borgman et al., 2007).

Research on resilient performance in relation to hazard therefore stands to benefit from a paradigmatic shift towards a research approach that engages the vast quantities of human-centered hazard data now being produced, necessitating a multi-method approach. As with any multi-method study—effectiveness "rests on the premise that the weaknesses in each single method will be compensated by the counter-balancing strengths of another" (Jick, 1979).

## Study Design and Evolution

### *Resilience Framework*

Early research in resilience engineering identified organizational factors thought to contribute to resilience (Woods, 2006), highlighting the importance of buffering capacity and flexibility (a concept similar to adaptive capacity) as organizations approach the edge of their operating envelope. For emergent or otherwise new organizations, assessment of the first two properties may be hamstrung by lack of operating history—even when the capabilities of individuals or subgroups within the organization are known or knowable. Subsequent research emphasizes the importance of learning from incidents and "normal" work (Saurin and Carim Júnior, 2011), and of system awareness (Costella et al., 2009; Hémond and Robert, 2012). As discussed elsewhere (Erol et al., 2010; Madni and Jackson, 2009), continued maturation of the field of resilience engineering depends critically on the development of tools and methodologies for measuring these and possibly other factors thought to contribute to resilience. For example, Erol et al. (2010) propose recovery time (that is, "the time taken for an enterprise to overcome disruption and return to its normal state") and level of recovery (that is, the operational state of the enterprise) as aspects of an enterprise's adaptive capacity.

Of course, the path to recovery of the enterprise towards the final level of recovery will also be of interest (that is, do recovery activities happen steadily, sporadically, periodically?).

Empirical indicants of these resilience measures vary widely, ranging from subjective assessments undertaken via questionnaires (Costella et al., 2009), qualitative rating scales (Hémond and Robert, 2012) and objective measures (Øien et al., 2010; Shirali et al., 2013). A distinguishing feature of much of this work is its emphasis on capturing the dynamics of complex systems (Balchanos et al., 2012).

The remainder of this section uses data from a single case study to investigate both the opportunities and limitations of process-level data (culled from various sources) to investigate organizational resilience, focusing primarily on the measurement of factors articulated by Woods (2006). It also considers some of the practical and theoretical difficulties involved in defining the structure of emergent organizations as well as the adaptive work they perform in seeking to achieve resilience.

*Case Study Background*

Following the 2001 World Trade Center attack, an organization comprised of individuals from both public- and private-sector organizations formed to manage debris removal activities at Ground Zero (Langewiesche, 2002)—the area immediately in and around the Twin Towers in New York City. Massive mobilizations of equipment and personnel were made, first to attempt to rescue any survivors and then to begin removal of material from the site to enable recovery of remains. It was soon obvious that new procedures and new organizations would be needed to conduct the operations (Langewiesche, 2002; Myers, 2003). As stated by Langewiesche (2002), "[t]he inapplicability of ordinary rules and procedures to such a chaotic environment required workers there to think for themselves, which they proved very capable of doing."

In the immediate aftermath of the event, rescue teams searched for possible survivors (Langewiesche, 2002), and mobilization of excavators and debris removal trucks began the evening of the attack (see Figure 11.1). In the next few days it became apparent that cranes would be needed to safely reach areas where the subgrade was too unstable for the excavators to traverse. Large cranes were brought in from various parts of the country in order to support the rescue of possible survivors, the recovery of remains, and the removal of debris (Langewiesche, 2002). The cranes "came in various sizes, from the 'small' 320s (which could pull apart an ordinary house in minutes) to the oversized 1200s, monstrous mining machines rarely seen in New York, which proved to be too awkward for many uses on the pile" (Langewiesche 2002). Once the location for the crane was specified, a mat or dunnage steel would be constructed on which the crane would set. Some situations required the construction of a crane ramp.

**Figure 11.1 Overview of site**

Site conditions created challenges in the placement of heavy equipment. An important concern was the integrity of the slurry wall (Tamaro, 2002), both because of concern it had been compromised by the towers' collapse, and because debris removal operations would subject it to unplanned-for stresses. For example, heavy equipment placed adjacent to the slurry wall "could cause the collapse of the slurry walls or any remaining basement structures [on the site]. A collapse of the slurry wall would mean inundation from the nearby Hudson River" (Tamaro, 2002).

The original objective of this work was to characterize (1) the operating capacity of the site (that is, the total loading and hauling capacity of the system), (2) the actual throughput of the site (that is, loads and tonnage hauled), and (3) organizational, technical and process-level factors which related operating capacity to actual performance. While this objective has not been achieved (for reasons discussed below), attempts to characterize the system along each of these three dimensions have yielded naturally to a discussion of prospects for assessing the organization's resilience post-hoc, and in light of Wood's framework (Woods, 2006).

From the start of this study, which began while cleanup operations were still ongoing, a fundamental tension was evident between achieving the objective of debris removal (or, perhaps more appropriately, site clearance) and ensuring site safety. Concurrently, organization structure and practices had to be developed to manage this tension. As described below, documentation of organizational and work structure and practice was therefore itself evolving—a situation likely to be found during other disaster response operations. Consistent with engineering practice, extensive records were kept of this "deconstruction" effort. For example, sometimes up to 20 printed copies of engineering drawings had to be delivered into the field. Document management services supported some tasks. For a brief period following conclusion of debris removal operations, many key documents were housed at a single location, thus facilitating the development of the work described here.

*Organizational Structure and Process*

An evolving, multi-organizational system undertook debris removal operations. New York City's Department of Design and Construction (DDC) coordinated the cleanup effort, with four construction companies—Turner, AMEC, Bovis and Tully—performing the actual demolition of the site across four respective sectors. A single organization was engaged to provide and coordinate all structural engineering services on a continuous basis, including assisting contractors with placement of equipment, designing support systems for cranes, damage assessment of neighboring buildings, and specifying stabilization of damaged buildings.

To support communication and documentation within and across shifts, the construction company responsible for a sector filed a daily, hand-written, paper-based *Field Report* at the end of each shift. Initially, there was usually one report per sector, although as time progressed one team might cover multiple sectors. Each report used a standard header containing the date of the meeting, the sector (usually denoted by the team name), and the names of the reporting engineers. The body of the report usually contained a number of observations, given as bulleted items, reporting the status of ongoing work and any issues for the next shift. A sample observation was "Received and Reviewed proposed demolition plan for west façade of Tower 2."

To provide insight into response behaviors (and thus into the organization's adaptive capacity), Field Reports were used to identify instances of decision-making and planning about debris removal equipment. An independent member of the research team was provided with all meeting notes and instructions for classifying the content of individual meeting items, reported on as "decision" or "plan" statements. Decision statements are statements about allocations of resources that had already been made. A plan statement referred to "a scheme, program, or method worked out beforehand for the accomplishment of an objective" or to "a proposed or tentative project or course of action." This category includes judgments (for example, "The company said they would remove the equipment by tomorrow."). A tally was also kept of reported work stoppages, along with the reasons for those stoppages, and of reports of inactive equipment.

### *On-site Risks and Operating Capacity*

As stated previously, the stability of the slurry wall encircling the site was an ongoing concern during cleanup activities. Potential sources of instability in the slurry wall were pressures exerted by cranes stationed on or near the wall itself, as well as the lateral pressure exerted by the land on the other side of the site of the wall. Raw data (in MS Excel format) were obtained on displacement measured at various locations on the slurry wall over time. These "shots" were taken from *stations* to various *points* on or near the slurry wall. The number of stations used during the study period is shown in Figure 11.2 (for example, shots were taken from 12 locations along Vesey St.). The readings from these shots were examined on a rolling basis: that is, engineers would examine recent readings for any meaningful deviation from a small number of prior readings. When the deviation was "too large," work in the area near the sensor would be stopped, further inspection performed, and any remedial actions undertaken.

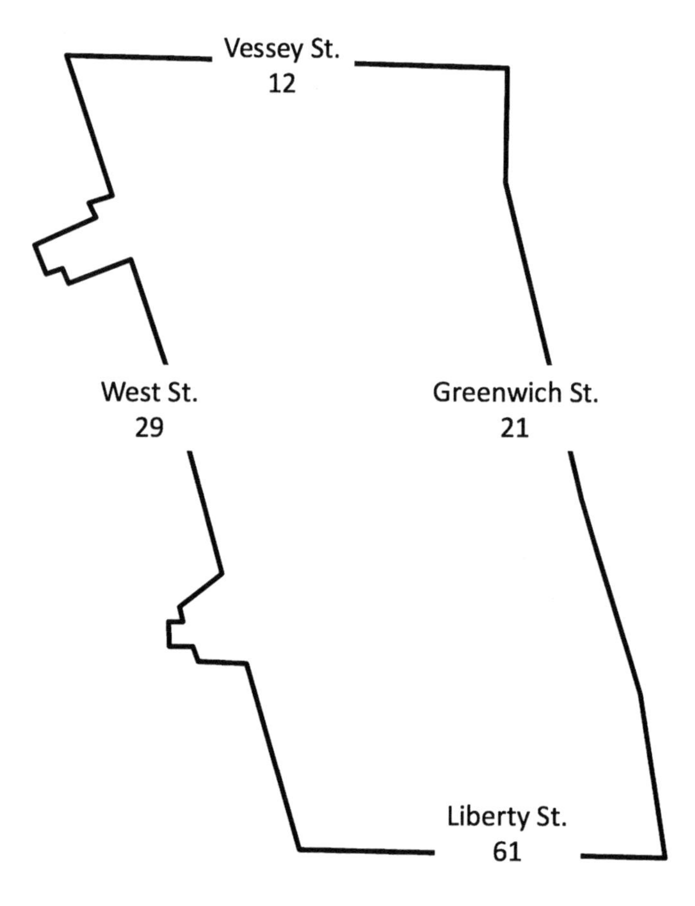


**Figure 11.2 Station location and counts**

The day-to-day locations of large debris removal equipment (that is, cranes) were recorded by field engineers who walked the site and annotated site plans. The annotations were then entered into AutoCAD drawings and archived as *Crane Maps*, which allowed engineers to view the location and approximate maximum reach of each crane. A sample Crane Map is shown in Figure 11.3. As new equipment made its way onto or from the site, or as cranes were moved around the site, the maps were updated. These maps were used to determine which cranes were present on-site during the study period, and to identify any movements of these cranes. Data on crane locations over time were extracted via semi-automated methods from the AutoCAD drawings.

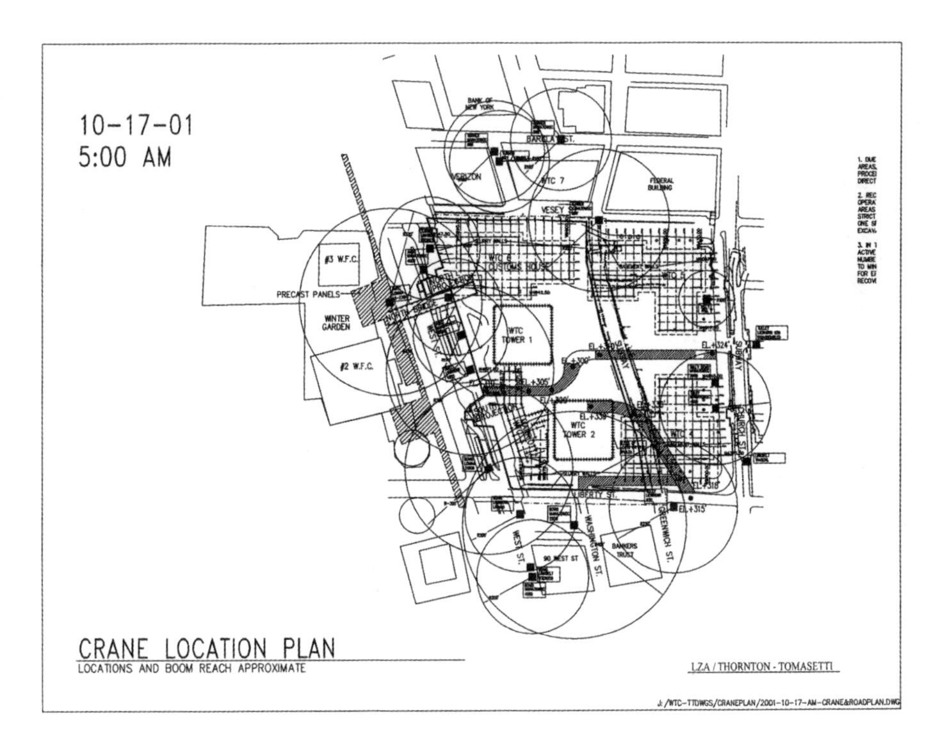


**Figure 11.3 Sample crane location plan**

Interviews with field personnel, however, strongly suggested that the bulk of debris removal work shifted over time to other types of equipment. Lighter, more agile equipment such as grapplers gradually took over much of the debris removal work. However, no records comparable to those of cranes were found within the official project archives. Various other sources were sought, until a private archive was found containing numerous photographs of the site, many of which were eventually included in the book *Aftermath* (Meyerowitz, 2006). A template was developed for using this photographic record to identify the location of grapplers and dump trucks. Additionally, the number of cranes in each field of view was recorded. A sample result from the application of this template is given in Figure 11.4. The figure shows the field of view of the indicated photo, counts for equipment, the date/time of the photo, and any comments.

*Organizational Performance*

Discussions with responsible personnel suggested that appropriate measures of *performance* included the amount of debris removed from the site, avoided human and economic losses, and the recovery of human remains. Data on debris removed from the site were obtained as the daily number and tonnage of loads leaving the site. These data were obviously collected at a much more disaggregate level (that is, that of the individual truck load) but were not available for analysis. Data on avoided human and economic losses (for example, as might be reflected in near-misses) appear not to have been collected directly, but are suggested in the Field Notes as work stoppages. Finally, data were also not available on the effectiveness of the recovery of human remains, despite the influence of this work on overall organizational performance. The recovery of human remains was an intermediary step between initial loading of material by grapplers or cranes, and transport of debris to a permanent landfill. Debris was loaded at a pickup point, hauled on-site to a second location, hand-sifted and inspected for human remains, then reloaded onto dump trucks and hauled off-site.

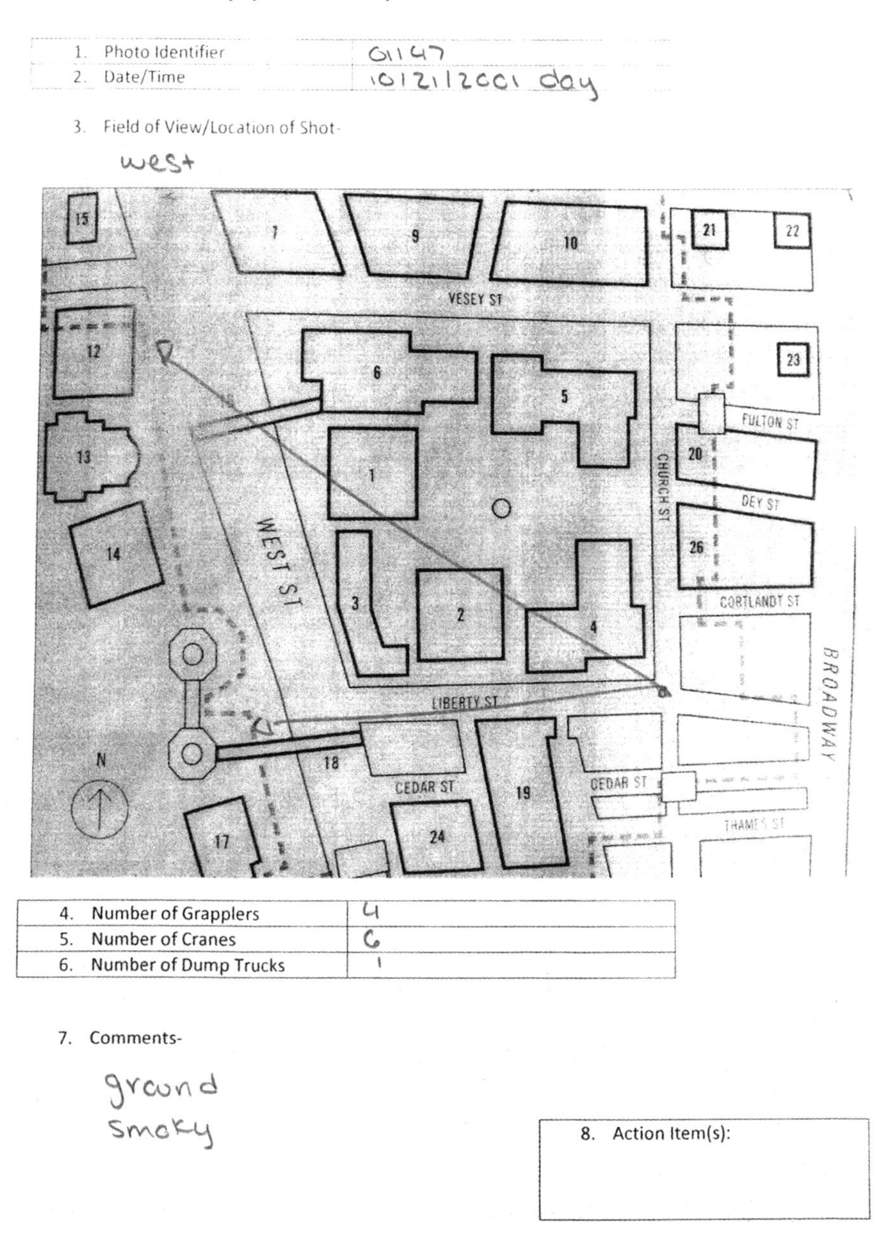

| 1. Photo Identifier | 01147 |
| --- | --- |
| 2. Date/Time | 10/21/2001 day |

3. Field of View/Location of Shot-

west



| 4. Number of Grapplers | 4 |
| --- | --- |
| 5. Number of Cranes | 6 |
| 6. Number of Dump Trucks | 1 |

7. Comments-

ground
smoky

8. Action Item(s):

**Figure 11.4 Coding template for equipment placement**

## Observations

The data discussed thus far offer a wealth of perspectives on the phenomena under study. However, they are organized on variable temporal and spatial scales. For example, time scales ranged from minutes (in the case of slurry wall readings) through multiple days (in the case of crane maps). Spatial scales, when present, ranged from fractions of an inch to feet. Indeed, the only common factor across all data was time—and only at the level of a day or multiple days. For purposes of illustration, the data are therefore presented at the scale of the individual day (see discussion below on missing data). The first set of study variables are summarized in Table 11.1.

**Table 11.1 Selected study variables**

| Series | Variable | Measurement | Source |
|---|---|---|---|
| 1 | Loads | Number of truckloads of debris removed from site | Debris reports |
| 2 | Tonnage | Tonnage of debris removed from site | Debris reports |
| 3 | # of Crane Movements | Number of times a crane was moved around, onto or off the site. | Crane maps |
| 4 | # of Cranes | Number of cranes on-site | Crane maps |
| 5 | Proportion of Active Equipment | Proportion of cranes that were in use out of total number of cranes | Field Reports |
| 6 | # of Stoppages | Reported stoppages in use of debris removal equipment | Field Reports |
| 7 | Proportion of # of Decisions | Proportion of reported decisions out of total number of decision and plan statements | Field Reports |

The values of these variables over time are given in Figure 11.5, as follows. The *first and second series* shows two key measures of performance—number of loads and tonnage removed from the site per day, respectively. Both measures show considerable fluctuation over the study period, with no obvious trends or periodicity.

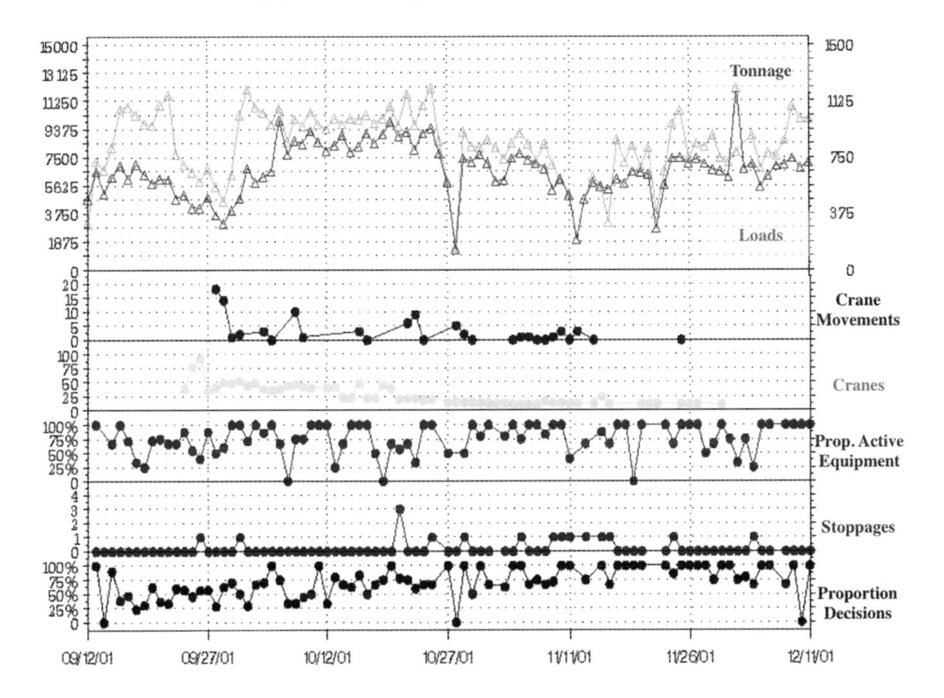


**Figure 11.5 Plots of selected study variables**

The *third series* shows the number of reported movements of cranes, either onto the site, off the site, or around the site. While data are not available for all days in the study period, it is possible to see that crane movements diminished over time, reflecting participants' reports that cranes became increasingly less effectual over the lifetime of the project, in part due to the increasing need to dispatch debris removal equipment to places that could not be reached by cranes and as access roads into the debris pile were constructed.

The reduction in the contribution of cranes to debris removal is also reflected in the *fourth series*, which shows the number of cranes on-site. The data from early in the study period show large numbers of cranes, while by late October the number had diminished by more than 75 per cent. On-site cranes remained active, however, as shown in the *fifth series*. All cranes were utilized for many of the days during the study period, at a rate that only occasionally declined to below 25 per cent. Using the results of the photo-coding procedure (see Figure 11.4) applied to photos from

30 days of this time period, an attempt was made to determine the number of grapplers and dump trucks on-site (photos of the site on other days were not available). Unfortunately, data were insufficient to enable a sufficient number of complete views of the site to be coded.

The declines shown in the fifth series are sometimes reflected in the *sixth series*, which shows the number of stoppages per day. Reported stoppages were infrequent until mid-October, and then averaged approximately once every two days for the remainder of the study period. In informal interviews with participants, it seems that not all short-term stoppages were reported in the Field Reports.

Finally, the *seventh series* shows the ratio of decision statement to planning statements in daily Field Reports. There is some suggestion of an overall trend towards proportionally greater reporting of decisions over plans as time progressed.

As discussed previously, some indication of on-site risk is provided by an examination of the measured displacements in the slurry wall. Sensors of various types were placed on the slurry wall slabs. Measurements were taken frequently earlier in the project, then with less regularity as the work progressed and it became clear that the slurry wall had stabilized. Displacement was measured along the north and south axes of the compass, as well as vertically. Because not all displacements were always collected, it is not possible to present a single value for displacement for each day. However, it is possible to consider the east-west, north-south and vertical displacements individually over the study period. The number of unique stations per day during the study period is shown in Figure 11.6, while the total number of shots taken per day is shown in Figure 11.7. Taken together, these figures clearly show considerable variability in measurement patterns over time. For example, early in the work, a relatively large number of measurements were taken from relatively small number of stations. Over time, the number of stations in use per day increased, but the average number of measurements per day remained relatively constant after early November.

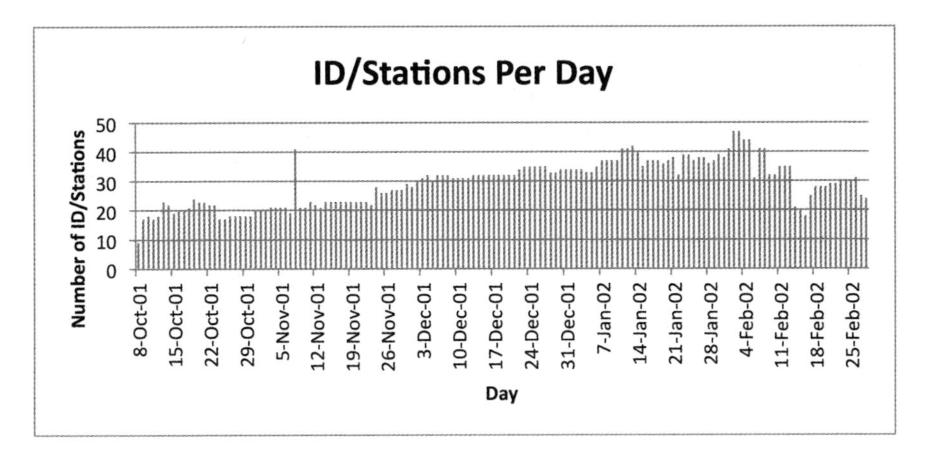


**Figure 11.6 Active stations per day**

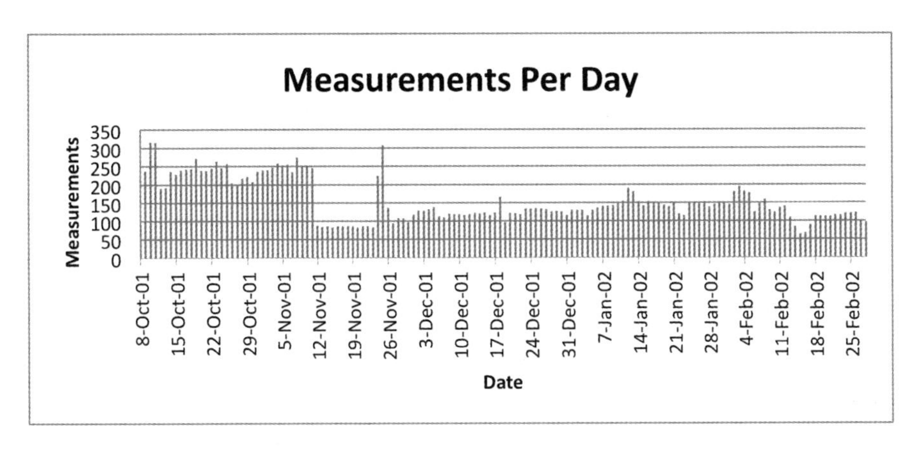


**Figure 11.7 Total measurements (i.e., shots) per day**

Performance as measured in loads is reasonably constant over time, with little evidence of periodicity or other trend over time. There is some suggestion that, after 27 October, overall performance diminished. This may of course in part be due to the increased difficulty in removing debris from the site. Over the course of the project, smaller capacity, more agile pieces of equipment were increasingly used, and these changes in the configuration of on-site equipment may have caused the organizations to rethink how to utilize heavy equipment. The increase in reports of decision-making over planning suggests that the organization became established in its procedures over time. Finally, some degree of risk of slurry wall collapse was often present, with more than 60 per cent of the observed displacements having a non-zero value.

A glaring change appears in late October, when reported load and tonnage totals decreased sharply. This change roughly coincided with an increased number of stoppages (series six). It may also be reflected in a drop in decision-making (that is, an increase in planning), as shown in the seventh series. The seventh series shows a gradual increase in the frequency of reported decisions relative to plans. The variable is computed as the number of statements reflecting decisions divided by the sum of this same number and the number of statements reflecting planned activities. In other words, the Field Reports are increasingly devoted to reporting of decisions rather than plans. Contextual factors such as work stoppages seem to have been handled in a way that did not result in diminishment of established capability for debris removal. However, the flow of debris from the site fluctuated considerably, and showed no sign of steady increases or decreases.

## Discussion and Conclusions

The organization responsible for debris removal at Ground Zero faced a well-defined task (complete clearance of the site) that was nonetheless both novel (in terms of scale and pace of activity) and risky (particularly with regard to the dangers associated with site flooding and with accidents associated with operation of debris removal equipment). A broad "systems" framework was guided data collection, which by necessity was opportunistic. Viewed through the lens of resilience engineering precepts (articulated some years later), the case offers a number of observations on prospects for the measurement of resilience within the framework specified by (Woods, 2006), as well as implications for the prospects for testing theories of organizational resilience.

A multidisciplinary perspective was necessary: potentially applicable data were embedded within multiple sources, and had to be extracted (and curated) through a highly variable range of instruments. Given the stated goals of the organization—site clearance without compromising safety—a systems perspective seemed appropriate. This perspective, while broad, nonetheless entailed a number of theoretical commitments, ranging from the necessity to measure on scales to the need to cast study variables in empirical relationships with each other. Expertise from various fields was brought to bear on the raw materials. For example, extensive consultations with civil engineers involved in the project were required in order to decode and explicate data on slurry wall displacement. Additionally, prior work on organizational decision-making informed the design of the instrument used for coding meeting minutes.

The original fundamental question of this research—how the organization balanced the demands of debris removal against safety—remains in some sense unanswered. The research strongly suggests that such systems-level questions must be posed early, and the data marshaled from the onset. But the data do reveal some of the dynamics underlying how buffering capacity and flexibility (particularly in organizational routines) may be brought to bear as organizations approach the edge of their operating envelope.

Buffering capacity, which may be viewed as the nature and extent of slack resources available relative to demands on those resources, is suggested by organizational performance (tonnage and loads hauled) relative to loading equipment available (here, cranes). Underlying this basic gap are decision-making processes related to resource use and planning, reflected in meeting notes as well as records of crane placements. A more extensive analysis might seek to associate planning processes with actual decision-making processes, particularly in light on on-site risk (for example, as reflected in slurry wall movements). A key observation here is that, in an open and evolving organization, both organizational resources and organizational performance targets are likely to co-evolve. Accordingly, the assessment of buffering capacity should engage dynamic modeling approaches that enable tracing of decision-making processes across related events (for example, from situation assessment through discussion of options and resource allocation, to processing of feedback from the field) (Butts, 2007).

As suggested by the foregoing discussion, research on organizational resilience depends critically on an ability to assess demands on the organization relative to its capacity to meet those demands, but also to determine the limits of organizational capabilities as demands exert considerable stress on organizational resources. When organizations are emergent or otherwise new, estimation of these capabilities may be based on the opinions of expert observers. The above research also suggests that pre-decisional processes such as planning (which may be tracked through analysis of appropriate documentation) may offer insights into the extent to which organizational capabilities are being taxed. To reiterate a previous example, the standard procedure of assigning cranes to debris removal duty was gradually replaced by one in which other equipment (such as grapples) were deployed, thus extending organizational capabilities to haul more loads. Because the performance envelope of an organization may not be known *a priori*, tools such as simulation may help enable members of the organization to ask "what if" questions as operations unfold. For example, it may be possible to postulate a model of risky choice that explains the link between sensor readings on the slurry wall to work stoppages and/or debris removal totals.

Studies of organizational resilience in relation to disaster frequently face the prospect of reconstructing response activities based on sparse and/or unreliable records. Often enough, recourse must be made to questionnaires in an attempt to develop "lessons learned." This study suggests how data produced during the response itself may be harnessed to yield observations on individual and organizational response activities, and to attempt to cast these observations in a systems-level framework.

In the future, distributed information technologies (such as networked handheld computers) will likely be of use in presenting management and operations personnel with historical data and, potentially, forecasts on-site conditions. A key consideration separating the design of such systems from the design of existing systems is that information requirements are not likely to be known with sufficient certainty a priori. In other words, information technologies to support decision-making must be capable of being reconfigured during the project to meet the needs of the project. A further challenge is raised by the increased presence of sensors on-site, and the ensuing questions of whether and how to integrate sensor and communications data to support the project.

## Acknowledgements

This work was supported in part by US National Science Foundation Grant CMS- 0301661. The author acknowledges the cooperation of the various organizations who provided assistance with this work, in particular the New York City Department of Design and Construction; Thornton Tomasetti, Inc.; Mueser Rutledge Consulting Engineers, Inc.; the US Federal Emergency Management Agency and the US Army Corps of Engineers. Valuable research assistance was provided by Louis Calabrese, Qing Gu, Arthur Hendela, Yoandi Interian, Rani Kalaria, and Jessica Ware.

## References

Balchanos, M., Y. Li, and D. Mavris. 2012. *Towards a Method for Assessing Resilience of Complex Dynamical Systems.* Paper to the *5th International Symposium on Resilient Control Systems (ISRCS),* Salt Lake City, UT, 15–18 August.

Borgman, C.L., J.C. Wallis, and N. Enyedy. 2007. Little science confronts the data deluge: Habitat ecology, embedded sensor networks, and digital libraries. *International Journal on Digital Libraries,* 7(1), 17–30.

Buchanan, D.A. and A. Bryman. 2007. Contextualizing methods choice in organizational research. *Organizational Research Methods,* 10(3), 483–7, 89–501.

Butts, C. 2007. Responder communication networks in the world trade center disaster: Implications for modeling of communication within emergency settings. *The Journal of Mathematical Sociology,* 31(2), 121–47.

Committee on Disaster Research in the Social Sciences: Future Challenges and Opportunities. (2006). Facing Hazards and Disasters: Understanding Human Dimensions. Washington, DC: The National Academies Press.

Costella, M.F., T.A. Saurin, and L.B. de Macedo Guimarães. 2009. A method for assessing health and safety management systems from the resilience engineering perspective. *Safety Science,* 47(8), 1056–67.

Dietze, M.C., D.S. LeBauer, and R. Kooper. forthcoming. On improving the communication between models and data. *Plant, Cell & Environment,*

Dombrowsky, W.R. 2002. Methodological changes and challenges in disaster research: Electronic media and the globalization of data collection, in *Methods of Disaster Research*, edited by R.A. Stallings. Philadelphia, PA: Xlibris Corporation, 305–19.

Erol, O., D. Henry, B. Sauser, and M. Mansouri. 2010. *Perspectives on Measuring Enterprise Resilience*. Paper to the *IEEE Systems Conference*, San Diego, CA, 5–8 April.

Hémond, Y. and B. Robert. 2012. Evaluation of state of resilience for a critical infrastructure in a context of interdependencies. *International Journal of Critical Infrastructures*, 8(2), 95–106.

Jick, T.D. 1979. Mixing qualitative and quantitative methods: Triangulation in action. *Administrative Science Quarterly*, 24 (Dec.), 602–59.

Langewiesche, W. 2002. *American Ground: Unbuilding the World Trade Center*. New York: North Point Press.

Madni, A.M. and S. Jackson. 2009. Towards a conceptual framework for resilience engineering. *IEEE Systems Journal*, 3(2), 181–91.

Meyerowitz, J. 2006. *Aftermath: World Trade Center Archive*.Phaidon.

Murthy, D. 2008. Digital ethnography: An examination of the use of new technologies for social research. *Sociology*, 42(5), 837–55.

Myers, M.F., ed. 2003. *Beyond September 11th: An Account of Post-disaster Research*. Boulder, CO: Natural Hazards Research and Applications Information Center, University of Colorado.

Øien, K., S. Massaiu, R. Tinmannsvik, F. Størseth. Development of Early Warning Indicators Based on Resilience Engineering. Paper to the *International Probabilistic Safety Assessment and Management Conference (PSAM10)* 7–11 June.

Saurin, T.A. and G.C. Carim Júnior. 2011. Evaluation and improvement of a method for assessing hsms from the resilience engineering perspective: A case study of an electricity distributor. *Safety Science*, 49(2), 355–68.

Savage, M. and R. Burrows. 2009. Some further reflections on the coming crisis of empirical sociology. *Sociology*, 43(4), 762–72.

Shirali, G.A., I. Mohammadfam, and V. Ebrahimipour. 2013. A new method for quantitative assessment of resilience engineering by pca and nt approach: A case study in a process industry. *Reliability Engineering & System Safety*, 119 (Nov.), 88–94.

Simmel, G. 1903. The metropolis and mental life. *The Urban Sociology Reader*, 23–31.

Summerfield, P. 1985. Mass-observation: Social research or social movement? *Journal of Contemporary History*, 20(3), 439–52.

Tamaro, G.J. 2002. World Trade Center 'Bathtub': From genesis to armageddon. *The Bridge*, 32(1), 11–17.

Vidaillet, B. 2001. Cognitive processes and decision making in a crisis situation: A case study, in *Organizational cognition: Computation and interpretation*, edited by T.K. Lant and Z. Shapira. Mahwah, NJ: Lawrence Erlbaum Associates, 241–63.

Weick, K.E. 1985. Systematic observational methods, in *Handbook of Social Psychology*, edited by G. Lindzey and E. Aronson. New York: Random House, 567–634.

Willcock, H.D. 1943. Mass-observation. *American Journal of Sociology*, 48(4), 445–56.

Woods, D. 2006. Essential characteristics of resilience, in *Resilience Engineering: Concepts and Precepts*, edited by E. Hollnagel, D. Woods, and N. Leveson. Aldershot, UK: Ashagate, 21–33.

# Chapter 12
# Becoming Resilient

Erik Hollnagel

## Introduction

The need to be safe, which in practice means the need to be able to operate or function with as many wanted and as few unwanted outcomes as possible, is paramount in every industry and indeed essential to all human activity (Hollnagel, 2014). It is obviously of particular concern in the many safety critical industries, where losing control of the process can lead to severe negative consequences. The examples of nuclear power generation and aviation immediately come to mind. But the need to be safe is present in all other industries as well, including healthcare and mining.

The pursuit of safety must obviously correspond to the nature of the systems and processes that are characteristic of an industry, in particular the ways in which things work and the ways in which they can fail. In the dawn of industrial safety – in the last half of the 18th century, when widespread use of steam engines heralded the second industrial revolution – the main problem was the technology itself (Hale and Hovden, 1998). The purpose of safety efforts was at first to avoid explosions and prevent structures from collapsing, and later to ensure that materials and constructions were reliable enough to provide the required functionality. After the end of the Second World War (1939–45), safety concerns were enlarged to include also the human factor. At first, human factors were about how people could be matched to technology and later about how technology could be matched to people. In both cases the main concern was productivity rather than safety. The full impact of the human factor on safety only became clear in 1979 following the accident at the nuclear power plant at Three Mile Island, which demonstrated how human activities and misunderstandings themselves could be safety critical. Following that, the notion of human error became so entrenched in both accident investigations and risk assessments that it often overshadowed other perspectives (Senders and Moray, 1991; Reason, 1990). In the late 1980s, safety concerns got a new focus when it was realised that organisational factors also were important. The triggering cases were the loss of the space shuttle Challenger and the accident at the nuclear power plant in Chernobyl, both in 1986. The latter also made safety culture an indispensable part of safety efforts (INSAG-1, 1986).

Looking back at these developments, the bottom line is that safety thinking must correspond to actual work in actual working conditions, that is, to what we can call the industrial reality. Simple methods, and simple models, may be appropriate for simple types of work and working environments, but cannot adequately account for what happens in more complicated work situations. The industrial reality is unfortunately not stable but seems forever to become more difficult to comprehend. So while safety efforts in the beginning of the 20th century only needed to concern themselves with technical systems, it is today necessary to address the set of mutually dependent socio-technical systems that is necessary to sustain a range of individual and collective human activities. Another important lesson is that the development of safety methods always lags behind the development of industrial systems and that progress usually takes place in jumps that are triggered by a major disaster. It would, however, clearly be an advantage if safety thinking could stay ahead of actual developments and if safety management could be proactive. Resilience engineering shows how to do that.

## Safety Culture

At the present time, industrial safety comprises different types of effort with different aims. Some focus on making the technology as reliable as possible. Others look to the human factor, both in the sense of the basic ergonomics of the workplace, and in the sense of creating suitable work processes and routines. Yet others focus on the organisational factor, in particular the safety culture. Indeed, safety culture has in many ways replaced 'human error' as the most critical challenge for safety management. This has created a widespread need to be able to do something about safety culture, which has not gone unheeded – despite the fact that safety culture as such remains an ill-defined concept.

The commonly used way of describing safety culture relies on a distinction between different levels, usually five (Parker, Lawrie and Hudson, 2006). The levels are treated as if they represent distinct expressions of safety culture although in practice they stand for representative positions on a continuum. A characterisation of the five levels is provided in Table 12.1. In recent writing, the generative level has been renamed the resilient level (Joy and Morrell, 2012), although the justification for that is not completely clear.

**Table 12.1 The five levels of safety culture**

| Level of safety culture | Characteristic | Typical response to incidents/accidents |
| --- | --- | --- |
| Generative (*resilient*) | Safe behaviour is fully integrated in everything the organisation does. | Thorough reappraisal of safety management policies and practices. |
| Proactive | We work on the problems that we still find. | Joint incident investigation. |
| Calculative | All necessary steps are followed blindly. | Regular incident follow-up. |
| Reactive | Safety is important, we do a lot every time we have an accident. | Limited investigation. |
| Pathological | The organisation cares more about not being caught than about safety. | No incident investigation. |

An important assumption underlying the idea of safety culture is that it always is in the organisation's interest to move to a higher level of safety culture. The motivation for wanting to improve the safety culture is, of course, that a degraded safety culture is assumed to be a major reason for the occurrence of accidents and incidents. But if we, for the sake of discussion, accept that assumption – noting in passing that it by no means has been independently proven – then it is only fair to acknowledge that there also are a number of arguments against making the journey. One is that it incurs a cost, and that it indeed may be rather expensive. A second is that moving from the reactive or calculative to the proactive or generative levels means that the organisation in addition to dealing with the certain, that which actually happens, also has to deal with the potential, that which could happen. This introduces a risk and requires that the organisation looks beyond the short term to the long term. That, however, does not fit the agenda of all organisations (Amalberti, 2013). In some cases it may be justified to remain calculative for a while, for instance to build up the resources that are necessary to survive in the long run. In that sense, the different levels of safety culture can be seen as corresponding to different priorities in the organisation's safety-related efficiency-thoroughness trade-off (Hollnagel, 2009).

### *About Safety Culture*

The first definition of safety culture was the result of the workshop at the International Atomic Energy Agency in 1986, in the wake of the accident at Chernobyl. Here safety culture was defined as '(t)hat assembly of characteristics and attitudes in organizations and individuals which establishes that, as an overriding priority, nuclear plant safety issues receive the attention warranted by their significance' (INSAG-1, 1986). This is quite similar to the definition of organisational culture as 'a pattern of shared basic assumptions invented, discovered, or developed by a given group as it learns to cope with its problems of external adaptation and internal integration' proposed by Edgar Schein in the 1980s (for example, Schein, 1992).

Since then a number of surveys of safety culture have tried to define the concept more precisely as well as to account for its role in safety, but with a rather disappointing outcome. Guldenmund (2000) began a review of theories and research on safety culture in a special issue of Safety Science as follows:

> In the last two decades empirical research on safety climate and safety culture has developed considerably but, unfortunately, theory has not been through a similar progression. … Most efforts have not progressed beyond the stage of face validity. Basically, this means that the concept still has not advanced beyond its first developmental stages.

Other surveys reach similar conclusions, for instance Hopkins (2006) and Choudhry, Fang and Mohamed (2007). Given this state of affairs, one could do worse than adopt the above mentioned IAEA definition of safety culture.

### *Becoming Safe*

The description of the five levels of safety culture almost forces people to think of safety culture development as a transition from one level or from one step of the ladder to the next (Parker, Lawrie and Hudson, 2006). The very description of the levels implies two things. First, that progress means moving up the levels, while deterioration means moving down. Second, that any change has to be done level-by-level (leaving out the possibility of a complete drop from one level to the bottom, of course).

In order to use safety culture developments to improve safety, we should be able to answer the following practical questions:

- First, *how can we determine* an organisation's current level of safety culture? This is important both to find out what the starting point is and when the goal or 'destination' has been reached. Otherwise a change might simply be completed when all the available resources – or the allotted time – have been spent. An answer to the first question must be unambiguous and operational. It must also refer to some kind of articulated theory rather than simply rely on established practice or social consensus. The answer is finally a prerequisite for being able to maintain a level, since maintenance requires that a change can be detected.
- Second, *what is the goal* in the sense of how good should the safety culture of the organisation be? Is the final goal that the organisation has a generative/resilient safety culture or could something less ambitious be acceptable? And what is a generative/resilient safety culture anyway, that is, how can

it be determined that an organisation has reached that level?
- Finally, *what kind of effort* does it take to change or move from one level to the next? Since the differences between the levels clearly are qualitative rather than quantitative, it is unlikely that a repetition of the same type of effort cumulatively will be sufficient to move an organisation from the lowest to the highest level. Related questions are how much effort is needed, how long time a change will take, whether the effects are immediate or delayed, how much effort it takes to remain at a given level, and what the 'distances' between the levels are – since it would be unreasonable to assume that the levels are equidistant.

A straightforward solution for how to improve safety would be to use a common development approach where change was guided or managed by an external agent (management). Change would in this way be the result of an active personal desire rather than a need to comply with management goals or requirements. In this way safety culture could be changed by 'winning hearts and minds' (Hudson, 2007).

While focusing on changes to individual behaviour may be consonant with the notion of organisational culture, it also means that individual performance and safety culture become entangled. It is therefore not really the organisation that moves from one level to the next, but rather the individuals that change their attitudes to their work in terms of personal responsibility, individual consequences and proactive interventions. Strictly speaking this means that a measure of the level of safety culture refers to some composite expression of the attitudes of the individuals rather than to an organisational characteristic. It also assumes that the organisation is homogeneous and that all its members share the same attitudes or the same culture. This clashes with the fact that all organisations are heterogeneous, which means that some parts (divisions, departments, special functions) may work in one way while others may work in another.

The final goal, that an organisation functions so that most things go right and few things go wrong, can also be achieved in other ways than by changing the safety culture. Instead of explaining performance as determined by the level of safety culture, we may look at the characteristics of organisational performance as such. This approach is used both by the High Reliability Organisation (HRO) school of thinking (Roberts, 1990; Weick, 1987) and resilience engineering (Hollnagel, Woods and Leveson, 2006; Hollnagel et al., 2011). In the following, we shall consider the implications of resilience engineering for understanding how safe performance can be brought about.

## Engineering a Culture of Resilience

The most important difference between resilience engineering and the common safety approaches is the focus on everyday successful performance. This is captured by the distinction between two safety concepts, called Safety-I and Safety-II (Hollnagel, 2014; ANSI, 2011). Safety-I defines safety as the absence of accidents and incidents, or as the 'freedom from unacceptable risk'. Safety-II defines safety as the ability to succeed under varying conditions. This is, of course, consistent with the hypothesis that a better safety culture leads to a reduction in incidents and accidents. But rather than propose an explanation based on a single factor or dimension, resilience engineering looks to the nature of individual and organisational performance. It is more precisely proposed that a resilient organisation – or system – must be able:

- to *respond* to regular and irregular variability, disturbances and opportunities;
- to *monitor* that which happens and recognise if something changes so much that it may affect the organisation's ability to carry out its current operations;
- to *learn* the right lessons from the right experience; and
- to *anticipate* developments that lie further into the future, beyond the range of current operations.

For any given organisation the proper mix or combination depends on the nature of its operations and the specific operating environment (business, regulatory, environmental, social and so on). The four abilities are furthermore mutually dependent. For instance, the effectiveness of an organisation's response to something depends on whether it is prepared (that is, able to monitor) and whether it has been able to learn from past experience.

The four abilities can clearly be developed to different degrees for a given organisation. And since each ability can be further specified by means of underlying or constituent functions, this can be used both to find how well an organisation currently *is* doing and to define how well it *should* be doing. A concrete approach to that is the Resilience Analysis Grid (RAG, cf., Hollnagel, 2010; ARPANSA, 2012), which can be used to assess how well an organisation performs on each of the four abilities at a given time. The same assessment can be used to propose concrete ways to develop specific sub-functions of an ability – without forgetting for a moment that the abilities are mutually dependent. The potential to develop the four abilities therefore offers a useful alternative to the safety journey, which in the following will be called the road to resilience.

In contrast to the development of safety culture, resilience engineering is not about reaching a certain level but about how well the organisation performs as such. Resilience does not characterise a state or a condition – what a system *is* – but how processes or performance are carried out – what an organisation *does*. Becoming resilient thus differs from becoming safe by being continuous rather than discrete. It is more precisely about maintaining a balance among the four abilities that is appropriate for a certain type of activity and a certain type of situation. If, for instance, an organisation focuses primarily on responding, as in handling unforeseen or difficult situations, and therefore neglects monitoring, then it is not considered to be resilient. The reason is simply that neglecting monitoring will increase the likelihood that performance is disturbed by unforeseen events, which will lead to reduced productivity and/or jeopardising safety.

## The Road to Resilience

In order to describe the road to resilience it is convenient to consider two extremes. One is a dysfunctional or pathological organisation that functions rather badly and the other a resilient organisation that functions well.

1. A dysfunctional organisation responds to what happens in a stereotypical or scrambled manner, and cares neither about monitoring, learning or anticipating. The ability to respond is fundamental since an organisation (or a system or an organism) that is unable to do so with reasonable effectiveness sooner or later will become extinct or 'die' – in some cases literally. An organisation that only is able to respond is reactive, and the absence of learning means that it constantly is taken by surprise.
2. A resilient organisation is able to respond, monitor, learn and anticipate. It is furthermore able to do all of these acceptably well, and to manage the required efforts and resources appropriately. But unlike the levels of safety culture, there is no ceiling for the abilities. It is always possible to respond more effectively or quickly, to improve the monitoring, to learn more and to anticipate better.

The practical question is how an organisation can change from being dysfunctional to become resilient. Since the four abilities all need to be improved, it would immediately seem as if there were at least four possible roads to resilience. The four 'roads' could be found by considering various combinations of the four resilience capabilities, and the transitions between them. By thinking a little more about the four abilities, it can be argued that one road to resilience is more sensible – and thereby also more effective – than the others.

### *The Dysfunctional Organisation*

The road to resilience in principle starts from the dysfunctional organisation, that is, an organisation that basically only is able to respond. While such an organisation is imaginable (such as a financial institution that is 'too large to fail' or a software company that does not recognise changes in the use of computing machinery), it cannot survive for long unless almost nothing happens around it. While a respond-only organisation is possible, the same does not go for the other three resilience abilities, for example, a monitoring-only, a learning-only or an anticipation-only organisation. The reason is simply that an organisation or a system cannot survive without the ability to respond, even if it only is opportunistically.

This leaves three possible ways forward, namely to enhance either the ability to monitor, to learn or to anticipate. While there are arguments for and against each of them, the overriding criterion for how to proceed should be that it enhances the organisation's ability to respond.

### *Organisations that Can Respond and Monitor*

The best way to enhance the ability to respond is to develop the ability to monitor. Monitoring allows an organisation to detect developments and disturbances before they become so large that a response is necessary. This will on the one hand enable the organisation to prepare itself for a response, for instance by reallocating internal resources or by changing its mode of operation, and on the other allow early responses to weak signals. A trivial example of that is proactive maintenance – which is better than scheduled maintenance and far better than emergency repairs. Responding at an early stage in the development of an event will generally require fewer resources and take less time, although it also incurs the risk that the response may be inappropriate or even unnecessary. Yet even that is preferable to operating in a reactive mode only.

As an example, consider an organisation where the conditions for everyday operation and production are unstable, and where there may be significant fluctuations in the supply of parts, of resources, in the quality of raw materials, in the environment and so on. In such cases it is important to keep an eye on the conditions that are necessary for safe and effective performance, hence to improve monitoring as the first step on the road to resilience.

### *Organisations that Can Respond, Monitor and Learn*

For an organisation that is able to respond and to monitor, the next logical step is to develop the ability to learn. Learning is necessary for several reasons. The most obvious is that the environment is changing, which means that there always will be new and unexpected situations or conditions. It is important to learn from these and to look for patterns and regularities that can improve the abilities to respond and to monitor. Another important reason is that the ability to respond always will be limited. It is simply not affordable to prepare a response for every event or for every possible set of conditions (Westrum, 2006). This means that the organisation every now and then will not know how to respond. It is clearly important to learn from these, to evaluate whether they are unique or likely to occur again, and to use that to improve both responding and monitoring. Similarly, the organisation should also learn from responses that went well, since it always is possible to make improvements. The organisation can use this experience to improve the precision of responses, the response time, the set of cues or indicators that are monitored and so on.

### *The Resilient Organisation*

At the point when an organisation can respond, monitor and learn sufficiently well, the ability to anticipate should be developed. Anticipation can more precisely be used to enhance the abilities to monitor (suggesting which indicators to look for), to respond (outlining possible future scenarios) and to learn (prioritising different lessons). Learning can be used to improve the ability to respond, to select appropriate indicators and cues and also to hone the imagination that provides the basis of anticipation. Monitoring can primarily be used to improve the ability to respond (increased readiness, preventive responses). And responding can provide the experience that is necessary to improve learning as well as anticipation.

This altogether means that an organisation that wants to improve, to become more resilient, must carefully choose how and when to develop each of the four abilities. It is not recommended to make a wholesale change or improvement, in the same way that the safety journey envisions a step change in the level of safety culture. An organisation must instead first determine how well it does with regard to each of the four abilities – by carefully assessing the functions that contribute to each ability – and then plan how to go about developing them. Such plans must take the mutual dependencies into account and look for the most efficient means rather than rely on ready-made solutions. In some cases an ability or its constituent functions may be developed by means of technical improvements, for example, better sensors or more powerful ways of analysing measurements. In others it may be the human factor or organisational relations that need working on. In yet other cases it may be organisational functions such as planning, event analysis and training. And there may finally be cases where attitudes, or even safety culture, are of instrumental value.

## Conclusions

The above considerations are summarised in Figure 12.1, which shows the main road to resilience. The road does not necessarily start by the dysfunctional organisation. Although such an organisation theoretically represents an extreme, it would not be able to survive for long in practice, and is therefore not a reasonable point of departure. The road rather starts from an organisation that is able to respond and monitor, that is, able to maintain an existence even if it is not the ideal.

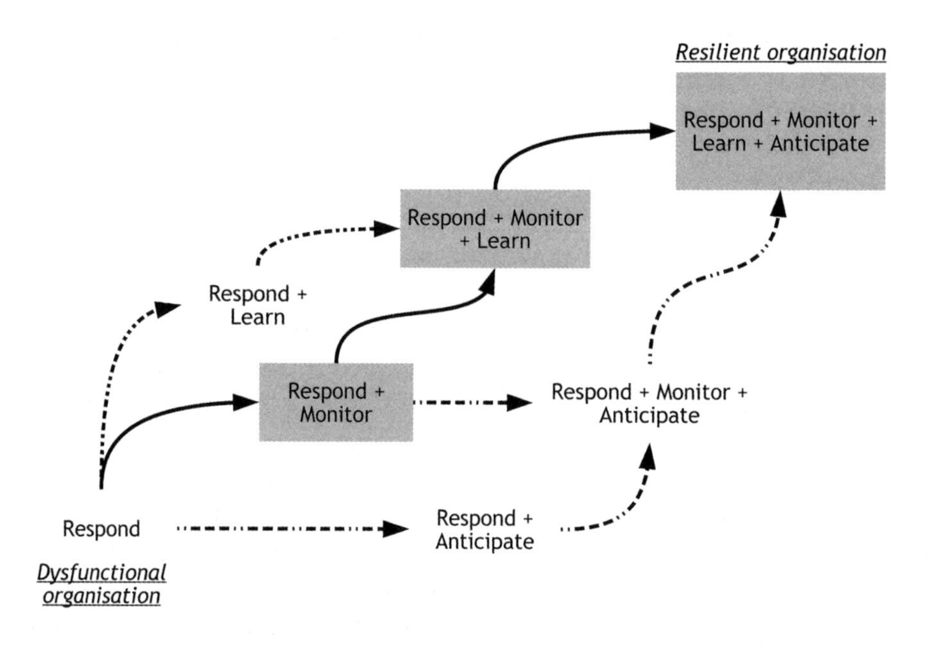


**Figure 12.1 The road to resilience**

The road ends by the resilient organisation, and the main path is shown by the dark lines in Figure 12.1. The progress is made by first developing the ability to learn followed by the ability to anticipate. But the organisation should also reinforce and improving already working abilities at the same time as it develops new ones. Progress is, however, neither simple or 'mechanical'. It requires an overall strategy as well as a 'constant sense of unease' (Nemeth et al., 2009) that can be used to make the best of the available means. The organisation's progress should be followed continuously so that any unanticipated development – or lack of development – can be caught early on and addressed operationally.

*The Three Questions and The Road to Resilience*

The consequences of the road to resilience can be summarised by looking at how the three questions were answered.

In terms of how can we determine the quality of the current performance of an organisation, resilience engineering proposes that this is done by looking at the four abilities and their underlying or constituent functions. There is already one practical technique for doing that, namely the Resilience Analysis Grid, which is derived from an articulated description of what resilience is.

In terms of setting the goal, resilience engineering does not prescribe a final solution. Instead, each organisation must decide how resilient its performance needs to be, expressed as differential levels of the four abilities. This is a pragmatic rather than a normative choice that depends heavily on what the organisation does and in which context it must work. Unlike the notion of safety culture as expressed by the five level model, there is no ceiling for resilience. An organisation can always improve and become better at what it does – in terms of productivity, safety, quality and so on.

Finally, with regard to how much effort it takes to improve resilience, the specification of the four abilities in terms of their constituent functions enables a high degree of realism. For each constituent function the possible means to make the desired change can be developed and evaluated in terms of cost, risk and so on. Since the constituent functions of the four abilities may differ significantly among functions, there is no standard or generic solution. But once the functions have been analysed and the goals defined, a variety of well-known and proven approaches will be available.

# Bibliography

## Christopher Nemeth


Adamski, A.J. and Westrum, R. (2003). Requisite Imagination: The fine art of anticipating what might go wrong. In E. Hollnagel (ed.), *Handbook of Cognitive Task Design*. 193–220.

International Council of Systems Engineers. Retrieved on line April 2013 from: http://www.incose.org/practice/whatissystemseng.aspx

Hollnagel, E. (2014). *Safety-I and Safety-II. The past and future of safety management*. Farnham, UK: Ashgate.

Hollnagel, E. and Woods, D.D. (2005). *Joint Cognitive Systems: Foundations of cognitive systems engineering*. Boca Raton, FL: Taylor and Francis/CRC Press.

Merriam Webster Dictionary. Retrieved online April 2013 from: http://www.merriam-webster.com/dictionary/engineering

Norman, D. 2011. *Living with Complexity*. Cambridge, MA: The MIT Press.

Reason, J. (1997). *Managing the Risks of Organizational Accidents*. Brookfield, VT: Ashgate Publishing.

Wreathall, J. and Merritt, A.C. (2003). Managing Human Performance in the Modern World: Developments in the US Nuclear Industry. In G. Edkins and P. Pfister (eds), *Innovation and Consolidation in Aviation*. Aldershot, UK: Ashgate Publishing.

Wreathall J. (2006). Properties of Resilient Organizations: An Initial View. In E. Hollnagel, D. Woods and N. Leveson (eds), *Resilience Engineering: Concepts and precepts*. Aldershot, UK: Ashgate Publishing. 275–85.

Woods, D.D. (2000, September). Behind Human Error: Human Factors Research to Improve Patient Safety. *National Summit on Medical Errors and Patient Safety Research*, Quality Interagency Coordination Task Force and Agency for Healthcare Research and Quality. http://www.apa.org/ppo/issues/shumfactors2.html


## Per Becker, Marcus Abrahamsson and Henrik Tehler


Abrahamsson, M., Hassel, H. and Tehler, H. (2010). Towards a system-oriented framework for analysing and evaluating emergency response. *Journal of Contingencies and Crisis Management*, 18(1), 14–25.

Beck, U. (1999). *World Risk Society*. Cambridge: Polity.

Belton, V. and Stewart, T.J. (2002). *Multiple Criteria Decision Analysis: An integrated approach*. Boston: Kluwer Academic Publishers.

Berkes, F. and Folke, C. (1998). Linking social and ecological systems for resilience and sustainability. In F. Berkes and C. Folke (eds), *Linking Social and Ecological Systems: Management practices and social mechanisms for building resilience*. Cambridge and New York: Cambridge University Press, 1–25.

CADRI (2011). *Basics of Capacity Development for Disaster Risk Reduction*. Geneva: Capacity for Disaster Reduction Initiative.

Cohen, L., Pooley, J.A., Ferguson, C. and Harms, C. (2011). Psychologists' understandings of resilience: Implications for the discipline of psychology and psychology practice. *Australian Community Psychologist*, 23(2), 7–22.

Cook, R.I. and Nemeth, C. (2006). Taking things in one's stride: Cognitive features of two resilient performances. In E. Hollnagel, D.D. Woods and N. Leveson (eds), *Resilience Engineering: Concepts and precepts*. Aldershot and Burlington: Ashgate.

Coppola, D.P. (2007). *Introduction to International Disaster Management*. Oxford: Butterworth-Heinemann (Elsevier).

Elsner, J.B., Kossin, J.P. and Jagger, T.H. (2008). The increasing intensity of the strongest tropical cyclones. *Nature*, 455(7209), 92–5.

Fordham, M.H. (2007). Disaster and development research and practice: A necessary eclecticism? In H. Rodríguez, E.L. Quarantelli and R.R. Dynes (eds), *Handbook of Disaster Research*. New York: Springer, 335–46.

Geist, H.J. and Lambin, E.F. (2004). Dynamic causal patterns of desertification. *Bioscience*, 54(9), 817–29.

Haimes, Y.Y. (1998). *Risk Modeling, Assessment, and Management*. New York and Chichester: John Wiley & Sons.

Haimes, Y.Y. (2004). *Risk Modeling, Assessment, and Management (2 ed.)*. Hoboken: Wiley-Interscience.

Hale, A. and Heijer, T. (2006). Is resilience really necessary? The case of railways. In E. Hollnagel, D.D. Woods and N. Leveson (eds), *Resilience Engineering: Concepts and precepts*. Aldershot and Burlington: Ashgate.

Haque, C.E. and Etkin, D. (2007). People and community as constituent parts of hazards: The significance of societal dimensions in hazards analysis. *Natural Hazards*, 41(41), 271–82.

Hewitt, K. (1983). The idea of calamity in a technocratic age. In K. Hewitt (ed.), *Interpretations of calamity*. London and Winchester: Allen & Unwin.

Hollnagel, E. (2006). Resilience– the challenge of the unstable. In E. Hollnagel, D.D. Woods and N. Leveson (eds), *Resilience Engineering: Concepts and precepts*. Aldershot and Burlington: Ashgate.

Hollnagel, E. (2009). The four cornerstones of resilience engineering. In C.P. Nemeth, E. Hollnagel and S. Dekker (eds), *Preparation and Restoration*. Farnham and Burlington: Ashgate, 117–33.

Kates, R.W., Clark, W.C., Corell, R., Hall, J.M., Jaeger, C.C., Lowe, I., et al. (2001). Sustainability science. *Science*, 292(5517), 641–2.

Leveson, N., Dulac, N., Zipkin, D., Cutcher-Gershenfeld, J., Carrol, J. and Barret, B. (2006). Engineering resilience into safety-critical systems. In E. Hollnagel, D.D. Woods and N. Leveson (eds), *Resilience Engineering: Concepts and precepts*. Aldershot and Burlington: Ashgate.

OECD (2003). Emerging systemic risks in the 21st century: An agenda for action. Paris: OECD.

Oliver-Smith, A. (1999). Peru's five-hundred-year earthquake: Vulnerability in historical context. In A. Oliver-Smith and S.M. Hoffman (eds), *The Angry Earth: Disaster in anthropological perspective*. London and New York: Routledge, –88.

Pariès, J. (2006). Complexity, emergence, resilience. In E. Hollnagel, D.D. Woods and N. Leveson (eds), *Resilience Engineering: Concepts and precepts*. Aldershot and Burlington: Ashgate.

Pendall, R., Foster, K.A. and Cowell, M. (2010). Resilience and regions: Building understanding of the metaphor. *Cambridge Journal of Regions, Economy and Society*, 3(1), 71–84.

Perrow, C. (1999a). *Normal Accidents: Living with high-risk technologies*. Princeton: Princeton University Press.

Perrow, C.B. (1999b). Organizing to reduce the vulnerabilities of complexity. *Journal of Contingencies and Crisis Management*, 7(3), 150–155.

Petersen, K.E. and Johansson, H. (2008). Designing resilient critical infrastructure systems using risk and vulnerability analysis. In E. Hollnagel, C.P. Nemeth and S. Dekker (eds), *Resilience Engineering Perspectives: Remaining sensitive to the possibility of failure*. Aldershot and Burlington: Ashgate.

Raco, M. (2007). Securing sustainable communities. *European Urban and Regional Studies*, 14(4), 305.

Rasmussen, J. (1985). The role of hierarchical knowledge representation in decisionmaking and system management. *IEEE Transactions on Systems, Man, and Cybernetics*, 15(2), 234–43.

Renn, O. (2008). *Risk Governance*. London and Sterling: Earthscan.

Schulz, K., Gustafsson, I. and Illes, E. (2005). *Manual for Capacity Development*. Stockholm: Sida.

Senge, P. (2006). *The Fifth Discipline: The art & practise of the learning organisation (2 ed.)*. London and New York: Currency & Doubleday.

Turner, B.L., Kasperson, R.E., Matson, P.A., McCarthy, J.J., Corell, R.W., Christensen, L., et al. (2003). A framework for vulnerability analysis in sustainability science. *Proceedings of the National Academy of Sciences of the United States of America*, 100(14), 8074–9.

Ulrich, W. (2000). Reflective practice in the civil society: The contribution of critical systems thinking. *Reflective Practice*, 1(2), 247–68.

Wisner, B., Blaikie, P.M., Cannon, T. and Davis, I. (2004). *At Risk: Natural hazards, people's vulnerability and disasters (2nd ed.)*. London: Routledge.

Yates, F.E. (1978). Complexity and the limits to knowledge. *American Journal of Physiology: Regulatory, Integrative and Comparative Physiology*, 4(235), R201–4.

## Jean Christophe Le Coze, Nicolas Herchin and Philippe Louys

Hollnagel E., Woods D.D, and Leveson, N. (2006). *Resilience Engineering: Concepts and precepts*. Ashgate.

Joerges, B. 1988. Large technical systems: Concepts and issues. In Mayntz, R., Hughes, T. 1988.(eds) *The Development of Large Technical Systems*. Schriften des Max-Planck-Instituts für Gesellschaftsforschung Köln, No. 2, ISBN 3-593-34032-1, Campus Verlag, Frankfurt/Main, New York

Klein, G. (2009). *Streetlights and Shadows. Searching for the keys to adaptive decision making*. The MIT Press.

La Porte, T. 1991. (ed) Social responses to large technical systems. Control or anticipation. Springer.

Le Coze, J.C. and Herchin, N. (2011). *Observing Resilience in Large Technical System. Third symposium on resilience engineering*. Juan les Pins.

Le Coze, J.C., Herchin, N. and Louys, P. (2012). *To Describe or to Prescribe? Working On Safety*. Sopot, Poland.

Le Coze, J.C. (2012). Towards a constructivist program in safety. *Safety Science*, 50, 1873–93.

Mayntz, R., Hughes, T. 1988.(eds) *The Development of Large Technical Systems*. Schriften des Max-Planck-Instituts für Gesellschaftsforschung Köln, No. 2, ISBN 3-593-34032-1, Campus Verlag, Frankfurt/Main, New York

Reason, J., 1990. *Human Error*. Cambridge University Press.

Simon, Herbert (1947). *Administrative Behavior*, (1947), New York, NY: Macmillan.

Sperandio, JC. 1977. La régulation des modes opératoires en fonction de la charge de travail chez les controlleurs de trafic aérien. le travail humain, 40, 249–256.

Weick, K., Sutcliffe, K.M. and Obstfeld, D. (1999). Organising for high reliability: processes of collective mindfullness, *Research in Organisational Behavior*, 21, 81–123.

## Robert L. Wears and L. Kendall Webb

Adamski, A.J. and Westrum, R. (2003). Requisite imagination: the fine art of anticipating what might go wrong. In E. Hollnagel (ed.), *Handbook of Cognitive Task Design*. Mahwah, NJ: Lawrence Erlbaum Associates, 193–220.

Cook, R.I. (2010, 2010). How Complex Systems Fail. Retrieved 19 September 2010, from http://www.ctlab.org/documents/Ch%2007.pdf

Dekker, S.W.A. (2011). *Drift into Failure: From hunting broken components to understanding complex systems*. Farnham, UK: Ashgate.

Dekker, S.W.A., Nyce, J. and Myers, D. (2012). The little engine who could not: 'rehabilitating' the individual in safety research. *Cognition, Technology & Work*, 1–6. doi: 10.1007/s10111-012-0228-5.

Hollnagel, E. (2009). *The ETTO Principle: Efficiency-Thoroughness Trade-off (Why Things That Go Right Sometimes Go Wrong)*. Farnham, UK: Ashgate.

Hollnagel, E. (2011). Prologue: the scope of resilience engineering. In E. Hollnagel, J. Pariès, D.D. Woods and J. Wreathall (eds), *Resilience Engineering in Practice: A guidebook* (pp. xxix–xxxiv). Farnham, UK: Ashgate.

Jackson, D., Thomas, M. and Millett, L.I. (eds) (2007). *Software for Dependable Systems: Sufficient evidence?* Washington, DC: National Academy Press.

Lanir, Z. (1986). Fundamental Surprises. Retrieved from: http://csel.eng.ohio-state.edu/courses/ise817/papers/Fundamental_Surprise1_final_copy.pdf

Marais, K.B. and Saleh, J.H. (2008). Conceptualizing and communicating organizational risk dynamics in the thoroughness-efficiency space. *Reliability Engineering & System Safety*, 93(11), 1710–1719.

March, J.G. (1991). Exploration and exploitation in organizational learning. *Organization Science*, 2(1), 71–87.

March, J.G., Sproull, L.S. and Tamuz, M. (1991). Learning from samples of one or fewer. *Organization Science*, 2(1), 1–13.

Maruyama, M. (1963). The second cybernetics: deviation-amplifying mutual causal processes. *American Scientist*, 5(2), 164–79.

Rochlin, G.I. (1999). Safe operation as a social construct. *Ergonomics*, 42(11), 1549–60.

Sagan, S.D. (1993). *The Limits of Safety: Organizations, accidents, and nuclear weapons*. Princeton, NJ: Princeton University Press.

Snook, S.A. (2000). *Friendly Fire: The accidental shoot-down of US Black Hawks over Northern Iraq*. Princeton, NJ: Princeton University Press.

Wears, R.L. (2010). Health information technology risks. *The Risks Digest*, 26(25). Retrieved from: http://catless.ncl.ac.uk/Risks/26.25.html#subj1

Wears, R.L., Cook, R.I. and Perry, S.J. (2006). Automation, interaction, complexity, and failure: a case study. *Reliability Engineering and System Safety*, 91(12), 1494–1501. doi: 10.1016/j.ress.2006.01.009.

Wears, R.L., Fairbanks, R.J. and Perry, S. (2012). *Separating Resilience and Success* (Proceedings of the Resilience in Healthcare). Middelfart, Denmark, 4–5 June 2012.

Wears, R.L. and Leveson, N.G. (2008). 'Safeware': safety-critical computing and healthcare information technology. In H.K and J.B. Battles, M.A. Keyes and M.L. Grady (eds), *Advances in Patient Safety: New Directions and Alternative Approaches* (AHRQ Publication No. 08-0034-4 ed., Vol. 4. Technology and Medication Safety, pp. 1–10). Rockville, MD: Agency for Healthcare Research and Quality.

Wears, R.L. and Morrison, J.B. (2013). *Levels of Resilience: Moving from resilience to resilience engineering* (Proceedings of the 5th International Symposium on Resilience Engineering (in review), Utrecht, the Netherlands, 25–27 June 2013.

Woods, D.D. and Branlat, M. (2011). Basic patterns in how adaptive systems fail. In E. Hollnagel, J. Paries, D.D. Woods and J. Wreathall (eds), *Resilience Engineering in Practice*. Farnham, UK: Ashgate, 127–44.

Woods, D.D. and Cook, R.I. (2006). Incidents–markers of resilience or brittleness? In E. Hollnagel, D.D. Woods and N. Levenson (eds), *Resilience Engineering*. Aldershot, UK: Ashgate, 70–76.

Woods, D.D., Dekker, S.W.A., Cook, R.I., Johannesen, L. and Sarter, N. (2010). *Behind Human Error* (2nd ed.). Farnham, UK: Ashgate.

Woods, D.D. and Wreathall, J. (2008). Stress-Strain Plots as a Basis for Assessing System Resilience. In E. Hollnagel, C.P. Nemeth and S.W.A. Dekker (eds), *Resilience Engineering: Remaining sensitive to the possibility of failure*. Aldershot, UK: Ashgate, 143–58.

## Masaharu Kitamura

Akaike, H. (1974). A new look at the statistical model identification. *IEEE Transactions on Automatic Control*, AC-19. 716–23.

Hatamura, Y. (Chairman) (2012). The Final Report of Investigation Committee on the Accident at Fukushima Nuclear Power Stations of TEPCO. Available at: http://www.cas.go.jp/jp/seisaku/ icanps/eng/final-report.html [accessed: 7 May 2013].

Hollnagel, E. (1993). *Human Reliability Analysis: Context and control*. London: Academic Press.

Hollnagel, E. (2006b), Epilogue of (Hollnagel, Woods and Leveson 2006).

Hollnagel, E. (2012), Resilience engineering at eight, Preface for Japanese translation of (Hollnagel, Woods and Leveson 2006).

Hollnagel, E. (2013). A tale of two safeties. *International Electronic Journal of Nuclear Safety and Simulation*, 4, 1–9.

Hollnagel, E., Woods, D.D., Leveson, N. (2006). *Resilience Engineering: Concepts and precepts*. Aldershot. UK: Ashgate Publishing.

Hollnagel, E., Paries, J, Woods, D.D. and Wreathall, J. (2011). *Resilience Engineering in Practice: A guidebook*. Aldershot. UK: Ashgate Publishing.

Kemeny, J.G. (Chairman) (1979). Report of The President's Commission on the Accident at the Three Mile Island. Available at: http://www.threemileisland.org/downloads/188.pdf [accessed: 28 March 2013].

Kitamura, M. (2009), The Mihama-2 accident from today's perspective, in E. Hollnagel, *Safer Complex Industrial Environments*. Boca Raton, FL: CRC Press, 19–36.

Kitazawa, K. (Chairman) (2012). The Fukushima Investigation Report by Independent Investigation Commission on the Fukushima Daiichi Nuclear Accident.

Klein, D. and Corradini, M. (Co-Chair) (2012). Fukushima Daiichi: ANS Committee Report. Available at: http://fukushima.ans.org/report/Fukushima_report.pdf [accessed: 3 April 2013].

Kurokawa, K. (Chairman) (2012). The Official Report of The National Diet of Japan by Fukushima Nuclear Accident Independent Investigation Commission. Available at: http://warp.da.ndl.go.jp/info:ndljp/pid/3856371/naiic.go.jp/en/report/ [accessed: 3 April 2013)

Reason, J. (2008). *The Human Contribution; Unsafe Acts, Accidents and Heroic Recoveries*. Aldershot UK: Ashgate.

Rissanen, J. (1978). Modeling by shortest data description. *Automatica*, 14 (5), 465–658.

Woods, D.D. and Cook,R.I. (2002). Nine steps to move forward from error. *Cognition, Technology and Work*, 4, 137–44.

Yagi, E., Takahashi, M. and Kitamura, M. (2006). Proposal of new scheme of science communication through repetitive dialogue forums, Proceedings of the 8th Probabilistic Safety Assessment and Management. New Orleans, USA, 14–18 May 2006.

## Tarcisio Abreu Saurin, Carlos Torres Formoso and Camila Campos Famá

Hollnagel, E. (2012). *FRAM: the Functional Resonance Analysis Method – modelling complex socio-technical systems*. Burlington: Ashgate.

Hollnagel, E. (2009). The four cornerstones of resilience engineering. In C. Nemeth, E. Hollnagel and S. Dekker (eds), *Resilience Engineering Perspectives: Preparation and restoration*, v. 2. Burlington: Ashgate, 177–33.

Hopkins, A. (2009). Thinking about process safety indicators. *Safety Science*, 47(4), 460–465.

Macchi, L. (2010). A resilience engineering approach to the evaluation of performance variability: development and application of the Functional Resonance Analysis Method for air traffic management safety assessment. Paris Institute of Technology, PhD thesis.

Neely, A., Richards, H., Mills, J. and Platts, K. (1997). Designing performance measures: a structured approach. *International Journal of Operations & Production Management*, 17(11), 1131–52.

Oien, K., Utne, I., Tinmannsvik, R. and Massaiu, S. (2011). Building safety indicators: part 2 – application, practices and results. *Safety Science*, 49, 162–71.

Saurin, T.A., Formoso, C.T. and Cambraia, F.B. (2008). An analysis of construction safety best practices from the cognitive systems engineering perspective. *Safety Science*, 46(8), 1169–83.

Wreathall, J. (2011). Monitoring – a critical ability in resilience engineering. In E. Hollnagel, J. Pariés, D.D. Woods and J. Wreathall (eds), *Resilience Engineering in Practice: a guidebook*. Burlington: Ashgate, 61–8.

## Amy Rankin, Jonas Lundberg and Rogier Woltjer

Cook, R.I., Render, M. and Woods, D.D. (2000). Gaps in the continuity of care and progress on patient safety. *BMJ* (Clinical research ed.), 320(7237), 791–4.

Cook, R.I. and Rasmussen, J. (2005). 'Going solid': a model of system dynamics and consequences for patient safety. *Quality & Safety in Health Care*, 14(2), 130–134.

Cook, R. and Woods, D.D. (1996) Adapting to new technology in the operating room. *Human Factors*, 38(4), 593–613.

Fischoff, B. (1975). Hindsight is not foresight: the effect of outcome knowledge on judgement under uncertainty. *Journal of Experimental Psychology: Human perception and performance*, 1(3), 288–99.

Furniss, D., Back, J., Blandford, A., Hildebrandt, M., & Broberg, H. (2011). A resilience markers framework for small teams. *Reliability Engineering & System Safety*, 96(1), 2–10.

Hoffman, R. and Woods, D.D. (2011). Beyond Simon's Slice: Five Fundamental Trade-Offs that Bound the Performance of Macrocognitive Work Systems. *IEEE Intelligent Systems*, 26(6), 67–71.

Hollnagel, E. (2008). The changing nature of risks. *Ergonomics Australia*, 22(1–2), 33–46.

Hollnagel, E. (2009). The Four Cornerstones of Resilience Engineering. In E. Hollnagel and S. Dekker (eds), *Resilience Engineering Perspectives, Vol 2 – Preparation and Restoration*. Farnham, UK: Ashgate, 117–33.

Hollnagel, E. (2012). Resilience engineering and the systemic view of safety at work: Why work-as- done is not the same as work-as-imagined. Bericht zum 58. Kongress der Gesellschaft für Arbeitswissenschaft vom 22 bis 24 Februar 2012. Dortmund: Gfa-Press, 19–24.

Hollnagel, E, and Woods, D.D. (2005). *Joint Cognitive Systems: Foundations of cognitive systems engineering*. Boca Ranton: CRC Press, Taylor & Francis Group.

Klein, G., Snowden, D. and Pin, C. (2010). Anticipatory thinking. In K. L. Mosier and U.M. Fischer (eds), *Informed by Knowledge: Expert performance in complex situations*. New York: Psychology Press.

Kontogiannis, T. (2009). A Contemporary View of Organizational Safety: Variability and Interactions of Organizational Processes. *Cognition, Technology & Work*, 12(4), 231–49.

Koopman, P. and Hoffman, R. (2003). 'Work-arounds, Make-work' and 'Kludges'. *IEEE Intelligent Systems* (Nov./Dec.), 70–75.

Loukopoulos, L.D., Dismukes, R.K. and Barshi, I. (2009). *The Multitasking Myth: Handling complexity in real-world operations*. Farnham, UK: Ashgate.

Lundberg, J. and Rankin, A. (2014). Resilience and vulnerability of small flexible crisis response teams: implications for training and preparation. *Cognition, Technology & Work*, 16(2), 143–155.

Lundberg, J., Törnqvist, E. and Nadjm-Tehrani, S. (2012). Resilience in Sensemaking and Control of Emergency Response. *International Journal of Emergency Management*, 8(2), 99–122.

Mumaw, R., Roth, E., Vicente, K. and Burns, C. (2000). There is More to Monitoring a Nuclear Power Plant than Meets the Eye. *Human Factors*, 42(1), 36–55.

Mumaw, R., Sarter, N., & Wickens, C. (2001). Analysis of Pilots Monitoring and Performance on an Automated Flight Deck. *In Proceedings of the 11th International Symposium on Aviation Psychology*. Colombus, OH.

Nemeth, C.P., Cook, R.I. and Woods, D.D. (2004). The Messy Details: Insights From the Study of Technical Work in Healthcare. *IEEE Transactions on Systems, Man, and Cybernetics – Part A: Systems and Humans*, 34(6), 689–92.

Nemeth, C.P., Nunnally, M., O'Connor, M.F., Brandwijk, M., Kowalsky, J. and Cook, R.I. (2007). Regularly irregular: how groups reconcile cross-cutting agendas and demand in healthcare. *Cognition, Technology & Work*, 9(3), 139–48.

Patterson, E.S., Roth, E.M., Woods, D.D., Chow, R. and Gomes, J.O. (2004). Handoff strategies in settings with high consequences for failure: lessons for health care operations. *International Journal for Quality in Health Care*, 16(2), 125–32.

Rankin, A., Dahlbäck, N. and Lundberg, J. (2013). A case study of factor influencing role improvisation in crisis response teams. *Cognition, Technology & Work*, 15(1), 79–93.

Rankin, A., Lundberg, J., Woltjer, R., Rollenhagen, C. and Hollnagel, E. (2014). Resilience in Everyday Operations: A Framework for Analyzing Adaptations in High-Risk Work. *Journal of Cognitive Engineering and Decision Making*, 8(1), 78–97.

Reason, J. (1997). *Managing the Risks of Organizational Accidents*. Burlington, VT: Ashgate.

Simon, H. (1969). *The Sciences of the Artificial*. Cambridge, MA: MIT Press.

Watts-Perotti, J. and Woods, D.D. (2007). How Anomaly Response Is Distributed Across Functionally Distinct Teams in Space Shuttle Mission Control Background: Overview of Anomaly Response. *Human Factors*, 1(4), 405–33.

Woods, D.D. (1993). The Price of Flexibility. *Knowledge-Based Systems*, 6, 1–8.

Woods, D.D. and Dekker, S.W.A. (2000). Anticipating the Effects of Technological Change: A New Era of Dynamics for Human Factors. *Theoretical Issues in Ergonomic Science*, 1(3), 272–82.

Woods, D.D., Dekker, S.W.A, Cook, R., Johannesen, L. and Sarter, N. (2010). *Behind Human Error (2nd ed.)*. Aldershot, UK: Ashgate.

## Akinori Komatsubara

Hollnagel, E. (1993). *Human Reliability Analysis: Context and Control*. UK: Academic Press.

Hollnagel, E. (2009). *The ETTO Principle: Efficiency-Thoroughness Trade-off, Why Things That Go Right Sometimes Go Wrong*. UK: Ashgate.

Hollnagel, E. (2011). Prologue: The Scope of Resilience Engineering, In E. Hollnagel, J.Paries, D.D. Woods and J. Wreathall (eds), *Resilience Engineering in Practice: A guidebook*. UK: Ashgate.

Hollnagel, E. (2012a). Resilience Health Care From Safety I to Safety II, *presentation slides at Resilient Health Care Network Tutorial, June 3, 2012.*

Hollnagel, E. (2012b). *FRAM: The Functional Resonance Analysis Method: Modeling complex socio-technical systems*, UK: Ashgate.

JTSB (2002). Japan Airlines Flight 907 and Japan Airlines Flight 958, A Near Midair Collision over the sea off Yaizu City, Shizuoka Prefecture, Japan at about 15:55 JST January 31, 2001, *Aircraft Accident Investigation Report No. 2003–5.*

Komatsubara, A. (2006). Human Defense-in-depth is Dependent on Culture. *Proceedings of the Second Resilience Engineering Symposium*, 165–72.

Komatsubara, A. (2008a). When Resilience does not work. In E. Hollnagel, C.P. Nemeth, S. Dekker (eds), *Remaining Sensitive to the Possibility of Failure.*UK: Ashgate, 79–90.

Komatsubara, A. (2008b). Encouraging People to do Resilience, *Proceedings of the Third Resilience Engineering Symposium 2008*, 141–7.

Komatsubara, A. (2011). Resilience Management System and Development of Resilience Capability on Site-Workers, *Proceedings of the fourth Resilience Engineering Symposium 2011* (pp. 148–54).

## Alexander Cedergren

Birkland, T.A. and Waterman, S. (2009). The Politics and Policy Challenges of Disaster Resilience. In C. P. Nemeth, E. Hollnagel and S. Dekker (eds), *Resilience Engineering Perspectives, Volume 2: Preparation and Restoration.* Farnham: Ashgate Publishing Limited, 15–38.

Cedergren, A. (2011). Challenges in Designing Resilient Socio-technical Systems: A Case Study of Railway Tunnel Projects. In E. Hollnagel, E. Rigaud and D. Besnard (eds), *Proceedings of the fourth Resilience Engineering Symposium.* Sophia Antipolis, France: Presses des MINES, 58–64.

Cedergren, A. (2013). Designing resilient infrastructure systems: a case study of decision-making challenges in railway tunnel projects. *Journal of Risk Research.* doi:10.1080/13669877.2012.726241.

De Bruijne, M., Boin, A. and Van Eeten, M. (2010). Resilience: Exploring the Concept and Its Meanings. In A. Boin, L.K. Comfort and C.C. Demchak (eds), *Designing Resilience: Preparing for extreme events.* Pittsburgh: University of Pittsburgh Press, 13–32.

Dekker, S. (2006). *The Field Guide to Understanding Human Error.* Aldershot: Ashgate Publishing Limited.

Hale, A. and Heijer, T. (2006). Is Resilience Really Necessary? The Case of Railways. In E. Hollnagel, D.D. Woods and N. Leveson (eds), *Resilience Engineering: Concepts and precepts.* Aldershot: Ashgate Publishing Limited, 125–47.

McDonald, N. (2006). Organizational Resilience and Industrial Risk. In E. Hollnagel, D.D. Woods and N. Leveson (eds), *Resilience Engineering: Concepts and precepts.* Aldershot: Ashgate Publishing Limited, 155–80.

Mendonça, D. (2008). Measures of Resilient Performance. In E. Hollnagel, C.P. Nemeth and S. Dekker (eds), *Resilience Engineering Perspectives: Remaining Sensitive to the Possibility of Failure* (Aldershot: Ashgate Publishing Limited, Vol. 1, 29–47.

Van Asselt, M.B.A. and Renn, O. (2011). Risk governance. *Journal of Risk Research*, 14(4), 431–49.

Vaughan, D. (1996). *The Challenger Launch Decision: Risky Technology, Culture, and Deviance at NASA*. Chicago: The University of Chicago Press.

Woods, D.D. (2003). Creating Foresight: How Resilience Engineering Can Transform NASA's Approach to Risky Decision Making. Testimony on The Future of NASA for Committee on Commerce, Science and Transportation. John McCain, Chair, October 29, 2003. Washington DC.

Woods, D.D. (2006). Essential Characteristics of Resilience. In E. Hollnagel, D.D. Woods and N. Leveson (eds), *Resilience Engineering: Concepts and precepts*. Aldershot: Ashgate Publishing Limited, 21–34.

Woods, D.D., Schenk, J. and Allen, T.T. (2009). An Initial Comparison of Selected Models of System Resilience. In C.P. Nemeth, E. Hollnagel and S. Dekker (eds), *Resilience Engineering Perspectives, Volume 2: Preparation and restoration*. Farnham: Ashgate Publishing Limited, 73–94.

## Johan Bergström, Eder Henriqson and Nicklas Dahlström

Airlines International. (April 2012). Training – man and machine, IATA, Accessed from http://www.iata.org/publications/airlines-international/april-2012/Pages/training.aspx on March 11, 2013.

Ashby, W.R. (1959). Requisite variety and its implications for the control of complex systems, *Cybernetica*, 1, 83–99.

Bergström, J. (2012). Escalation: Explorative studies of high-risk situations from the theoretical perspectives of complexity and joint cognitive systems. Doctoral dissertation, Lund: Media-Tryck.

Bergström, J., Dekker, S.W.A., Nyce, J. M. and Amer-Wåhlin, I. (2012). The social process of escalation: A promising focus for crisis management research. *BMC Health Services Research*, 12(1), 161. doi:10.1186/1472-6963-12-16.

Bergström, J., Dahlström, N., Dekker, S.W.A. and Petersen, K. (2011). Training organizational resilience in escalating situations. In E. Hollnagel, J. Pariès, D.D. Woods and J. Wreathall (eds), *Resilience Engineering in Practice: A guidebook*. Farnham, Surrey, England: Ashgate Publishing Limited, 45–57.

Bergström, J., Dahlström, N., Henriqson, E. and Dekker, S.W.A. (2010). Team coordination in escalating situations: an empirical study using mid-fidelity simulation. *Journal of Contingencies and Crisis Management*, 18(4).

Cilliers P. (1998). *Complexity and Postmodernism: Understanding complex systems*. London: Routledge.

Cilliers, P. (2005). Complexity, Deconstruction and Relativism. *Theory Culture &Society*, 22(5), 255–67.

Cook, R.I., Render, M. and Woods, D.D. (2000) Gaps in the continuity of care and progress on patient safety. *British Medical Journal*, 320:7237, 791–4.

Dahlström, N., Dekker, S. W. A., van Winsen, R., & Nyce, J. (2009). Fidelity and validity of simulator training. *Theoretical Issues in Ergonomics Science*, 10(4), 305–314. doi:10.1080/14639220802368864

Dekker, S.W.A. (2005) *Ten Questions about Human Error*. Aldershot, UK: Ashgate.

Dekker, S.W.A. (2011). *Drift into Failure: From hunting broken components to understanding complex systems*. Aldershot, UK: Ashgate.

Dekker, S., Bergström, J., Amer-Wåhlin, I. and Cilliers, P. (2012). Complicated, complex, and compliant: Best practice in obstetrics. *Cognition, Technology & Work*. doi:10.1007/s10111-011-0211-6.

European Aviation Safety Agency (2012). Terms of Reference – ToR RMT.0411 (OPS.094) Issue 2 Cologne, Germany: EASA.

Flin, R., O'Connor, P. and Crichton, M. (2008). *Safety at the Sharp End, A guide to non-technical skills*. Aldershot: Ashgate Publishing Company.

Henriqson, E., Saurin, T.A. and Bergström, J. (2010). Coordination as distributed and situated cognitive phenomena in aircraft cockpits. *Aviation in Focus*, 01, 58–76.

Hollnagel, E. (2011). The Scope of Resilience Engineering. In E. Hollnagel, J. Pariès, D.D. Woods and J. Wreathall (eds), *Resilience Engineering in Practice: A guidebook*. Farnham, Surrey, England: Ashgate Publishing Limited, xxix–xxxix.

Hollnagel, E. and Woods, D.D. (2005). *Joint Cognitive Systems: Foundations of cognitive systems engineering*. Boca Raton, FL: Taylor & Francis.

Hutchins, E. (1995a). How a cockpit remembers its speeds. *Cognitive Science*, 19, 265–88.

Hutchins, E. (1995b). *Cognition in the Wild*. Cambridge, MA: MIT Press.

Klein, G., Feltovich, P.J., Bradshaw, J.M. and Woods, D.D. (2005). Common ground and coordination in joint activity. In W. Rouse and K. Boff (eds), *Organizational Simulation*. New York: John Wiley & Sons.

Neisser, U. (1976). *Cognition and Reality*. San Francisco: W.H. Freeman.

Nyssen, A.S. (2011). From Myopic Coordination to Resilience in Socio-technical Systems: A Case Study in a Hospital. In E. Hollnagel, J. Pariès, D.D. Woods and J. Wreathall (eds), *Resilience Engineering in Practice: A guidebook*. Farnham, Surrey, England: Ashgate Publishing Limited, 219–35.

Palmqvist, H., Bergström, J. and Henriqson, E. (2011). How to assess team performance in terms of control: A cognitive systems engineering approach. *Cognition Technology & Work*, 14(4), 337–53. doi:10.1007/s10111-011-0183-6.

UK Civil Aviation Authority (2012). Monitoring Matters - Guidance on the Development of Pilot Monitoring Skills. CAA Paper 2013/02. West Sussex UK: UK CAA.

Voss, W. (2012). Evidence-Based Training. *AeroSafety World Magazine*, Flight Safety Foundation, accessed from: http://flightsafety.org/aerosafety-world-magazine/nov-2012/evidence-based-training on 11 March 2013.

Woods, D.D. (2003). Discovering how distributed cognitive systems work. In E. Hollnagel (ed.), *Handbook of Cognitive Task Design*. Hillsdale, NJ: Lawrence Erlbaum Associates, 37–54.

## Elizabeth Lay and Matthieu Branlat

Cook, R. (2012) Why resilience matters? Presentation at University of BC School of Population and Public Health Learning Lab, 7 May 2012.

Cook, R.I. and Nemeth, C. (2006). Taking Things in One's Stride: Cognitive Features of Two Resilient Performances. In E. Hollnagel, D.D. Woods and N. Leveson (eds), *Resilience Engineering: Concepts and precepts*. Aldershot, UK: Ashgate, 205–21.

De Meyer, A., Loch, C.H. and Pich, M.T. (2002). Managing Project Uncertainty: From Variation to Chaos, *MIT Sloan Management Review*, Winter Vol., 60–67.

Hollnagel, E. (2009). *The ETTO Principle: Efficiency-Thoroughness Trade-Off-Why Things That Go Right Sometimes Go Wrong*. Farnham, UK: Ashgate.

Hollnagel, E. (2012). How do we recognize resilience? Presentation at University of BC School of Population and Public Health Learning Lab, 7 May 2012.

Hong, L. and Page, S. (2004). Groups of diverse problem solvers can outperform groups of high-ability problem solvers. *PNAS*, 101(46), 16385–9.

Lay, E. (2011). Practices for Noticing and Dealing with the Critical. A Case Study from Maintenance of Power Plants. In E. Hollnagel, J. Pariès, D.D. Woods and J. Wreathall (eds), *Resilience Engineering in Practice*. Farnham, UK: Ashgate, 127–44.

Platt, M.L. and Huettel, S.A. (2008). Risky business: the neuroeconomics of decision making under uncertainty. *Nature Neuroscience*, 11, 398–403.

Taleb, N.N. (2010). *The Black Swan. 2nd edition*. London: Random House.

Weick, K.E. and Sutcliffe, K.M. (2001). *Managing the Unexpected: Assuring high performance in an age of complexity (1st ed.)*. San Francisco, CA: Jossey-Bass.

Westrum, R. (1993). Thinking by groups, organizations, and networks: A sociologist's view of the social psychology of science and technology. In W. Shadish and S. Fuller (eds), *The Social Psychology of Science*. New York: Guilford, 329–32.

Woods, D.D. (2006). Essential characteristics of resilience. In E. Hollnagel, D.D. Woods and N. Leveson (eds), *Resilience Engineering: Concepts and precepts*. Aldershot, UK: Ashgate, 19–30.

Woods, D.D. and Branlat, M. (2010). Hollnagel's test: being 'in control' of highly interdependent multi- layered networked systems. *Cognition, Technology & Work*, 12(2), 95–101.

Woods, D.D. and Branlat, M. (2011). Basic Patterns in How Adaptive Systems Fail. In E. Hollnagel, J. Pariès, D.D. Woods and J. Wreathall (eds), *Resilience Engineering in Practice*. Farnham, UK: Ashgate, 127–44.

Woods, D.D. and Wreathall, J. (2008). Stress-strain Plot as a Basis for Addressing System Resilience. In E. Hollnagel, C.P. Nemeth and S.W.A. Dekker (eds), *Resilience Engineering Perspectives: Remaining sensitive to the possibility of failure*. Adelshot, UK: Ashgate, 143–58.

Wreathall, J. and Woods, D.D. (2008). *The Stress-Strain Analogy of Organizational Resilience. Remaining Sensitive to the Possibility of Failure.* In E. Hollnagel, C. Nemeth and S. Dekker, *Resilience Engineering Perspectives: Remaining sensitive to the possibility of failure*. Burlington, VT, Ashgate Publishing Co.

## David Mendonça

Borgman C.L., Wallis J.C. and Enyedy N. (2007). Little Science Confronts the Data Deluge: Habitat Ecology, Embedded Sensor Networks, and Digital Libraries. *International Journal on Digital Libraries*, 7(1), 17–30.

Bryman A. (2012). *Social Research Methods*. Oxford, UK: Oxford University Press. Buchanan DA.

Bryman A. (2007). Contextualizing Methods Choice in Organizational Research. *Organizational Research Methods*, 10(3), 483–7, 89–501.

Butts C. (2007). Responder Communication Networks in the World Trade Center Disaster: Implications for Modeling of Communication within Emergency Settings. *The Journal of Mathematical Sociology*, 31(2), 121–47.

Committee on Disaster Research in the Social Sciences: Future Challenges and Opportunities. (2006). Facing Hazards and Disasters: Understanding Human Dimensions. Washington, DC: The National Academies Press.

Dietze M.C., LeBauer D.S. and Kooper R. (2013). On Improving the Communication between Models and Data. *Plant, Cell & Environment*, forthcoming.

Dombrowsky W.R. (2002). Methodological Changes and Challenges in Disaster Research: Electronic Media and the Globalization of Data Collection'. In R.A. Stallings (ed.), *Methods of Disaster Research*. Philadelphia, PA: Xlibris Corporation, 302–19.

Drabek, T.E. (2002). Responding to High Water: Social Maps of Two Disaster-Induced Emergent Multiorganizational Networks. Western Social Science Association, Albuquerque, NM, 2002.

Drabek T.E. and Haas, J.E. (1969). Laboratory Simulation of Organizational Stress. *American Sociological Review*, 34(2), 223–38.

Drabek T.E. and McEntire D.A. (2002). Emergent Phenomena and Multiorganizational Coordination in Disasters: Lessons from the Research Literature. *International Journal of Mass Emergencies and Disasters*, 20(2), 197–224.

Jick, T.D. (1979). Mixing Qualitative and Quantitative Methods: Triangulation in Action. *Administrative Science Quarterly*, 24(Dec.), 602–59.

Klein H.K. and Myers, M.D. (1999). A Set of Principles for Conducting and Evaluating Interpretive Field Studies in Information Systems. *MIS Quarterly*, 23(1), 67–94.

Langewiesche, W. (2002). *American Ground: Unbuilding the World Trade Center*. New York.

Meyerowitz, J. (2006). *Aftermath: World Trade Center Archive*. Phaidon.

Murthy, D. (2008). Digital Ethnography: An Examination of the Use of New Technologies for Social Research. *Sociology*, 42(5), 837–55.

Myers M.F. (ed.) (2003). Beyond September 11th: An Account of Post-Disaster Research. Boulder, CO: Natural Hazards Research and Applications Information Center, University of Colorado.

Orlikowski W.J. and Baroudi, J.J. (1991). Studying Information Technology in Organizations: Research Approaches and Assumptions. *Information Systems Research*, 2(1), 1–28.

Quarantelli, E.L. (1996). Emergent Behavior and Groups in the Crisis Time of Disasters. In K.M. Kwan (ed.), *Individuality and Social Control: Essays in Honor of Tamotsu Shibutani*. Greenwich, CT: JAI Press, 47–68.

Savage, M. and Burrows, R. (2007). 'The Coming Crisis of Empirical Sociology. *Sociology*, 41(5), 885–99.

Savage, M. and Burrows, R. (2009). Some Further Reflections on the Coming Crisis of Empirical Sociology. *Sociology*, 43(4), 762–72.

Simmel, G. (1903). The Metropolis and Mental Life. *The Urban Sociology Reader*, 23–31.

Summerfield, P. (1985). Mass-Observation: Social Research or Social Movement? *Journal of Contemporary History*, 20(3), 439–52.

Tamaro, G.J. (2002). World Trade Center 'Bathtub': From Genesis to Armageddon. *The Bridge*, 32(1), 11–17.

Tierney, K.J. (2007). From the Margins to the Mainstream? Disaster Research at the Crossroads. *Annual Review of Sociology*, 33(1), 503–25.

Vidaillet, B. (2001). Cognitive Processes and Decision Making in a Crisis Situation: A Case Study. In TK Lant, Z Shapira (eds), *Organizational Cognition: Computation and interpretation*. Mahwah, NJ: Lawrence Erlbaum Associates, 241–63.

Walsham, G. (1995). Interpretive Case Studies in Is Research: Nature and Method. *European Journal of Information Systems*, 4(2), 74–81.

Weick, K.E. (1985). Systematic Observational Methods. In G. Lindzey and E. Aronson (eds), *Handbook of Social Psychology*. New York: Random House, 567–634.

Welbank, M. (1990). An Overview of Knowledge Acquisition Methods. *Interacting with Computers*, 2(1), 83–91.

Willcock, H.D. (1943). Mass-Observation. *American Journal of Sociology*, 48(4), 445–56.

Wright, G. and Ayton, P. (1987). Eliciting and Modelling Expert Knowledge. *Decision Support Systems*, 3(1), 13–26.

York, R. and Clark, B. (2006). Marxism, Positivism, and Scientific Sociology: Social Gravity and Historicity. *Sociological Quarterly*, 47(3), 425–50.

## Erik Hollnagel

Amalberti, R. (2013). *Navigating safety. Necessary compromises and trade-offs – Theory and practice*. Dordrecht: Springer Verlag.

American National Standards Institute (ANSI). (2011). Prevention through design. Guidelines for addressing occupational hazards and risks in design and redesign processes (ANSI/ASSE Z590.3 – 2011). Des Plaines, IL: American Society of Safety Engineers.

Australian Radiation Protection and Nuclear Safety Agency (ARPANSA). (2012). Holistic safety guidelines, v1 (OS-LA-SUP-240U). Melbourne, Australia.

Choudhry, R.M., Fang, D. and Mohamed, S. (2007). The nature of safety culture: A survey of the state-of-the-art. *Safety Science*, 45(10), 993–1021.

Guldenmund, F.W. (2000). The nature of safety culture: A review of theory and research. *Safety Science*, 34, 215–57.

Hale, A.R. and Hovden, J. (1998). Management and culture: The third age of safety. In A-M. Feyer and A. Williamson (eds), *Occupational Injury: Risk, Prevention and Intervention*. London: Taylor and Francis, 129–66.

Hollnagel, E. (2009). *The ETTO principle: Efficiency-Thoroughness Trade-Off. Why things that go right sometimes go wrong*. Farnham, UK: Ashgate.

Hollnagel, E. (2011). Epilogue: RAG – The resilience analysis grid. In E. Hollnagel et al. (eds), *Resilience Engineering in Practice: A guidebook*. Farnham, UK: Ashgate.

'Hollnagel, E. (2014). *Safety-I and Safety-II. The past and future of safety management*. Farnham, UK: Ashgate.

Hollnagel, E. et al. (eds) (2011). *Resilience Engineering in Practice: A guidebook*. Farnham, UK: Ashgate.

Hollnagel, E., Woods, D.D. and Leveson, N. (eds) (2006). *Resilience Engineering: Concepts and precepts*. Farnham, UK: Ashgate.

Hopkins, A. (2006). Studying organisational cultures and their effects on safety. *Safety Science*, 44, 875–89.

Hudson, P. (2007). Implementing a safety culture in a major multi-national. *Safety Science*, 45, 697–722.

INSAG-1 (1986). Summary Report on the Post-Accident Review Meeting on the Chernobyl Accident Report by the International Nuclear Safety Advisory Group (STI/PUB/740). Wien: International Atomic Energy Agency.

Joy, J. and Morrell, A. (2012). *Operational Management Processes Manual*. University of Queensland: SMI Minerals Industry Safety and Health Centre.

Nemeth, C.P., Hollnagel, E. and Dekker, S. (eds) (2009). *Resilience Engineering Perspectives, Volume 2. Preparation and restoration*. Farnham, UK: Ashgate.

Parker, D., Lawrie, L. and Hudson, P. (2006). A framework for understanding the development of organisational safety culture. *Safety Science*, 44, 551–62.

Reason, J.T. (1990). *Human Error*. Cambridge: Cambridge University Press.

Roberts, K.H. (1990). Some characteristics of one type of high-reliability organization. *Organization Science*, 2, 160–176.

# Author Index

# Subject Index